TALES *from an* URBAN WILDERNESS

TALES *from an* URBAN WILDERNESS

BY SCOTT HOLINGUE
WITH KENAN HEISE

PREFACE BY ROBERT CROMIE

Chicago Historical Bookworks
831 Main St.
Evanston, Il. 60202

1994

10 9 8 7 6 5 4 3 2 1

ISBN: 0-924772-25-5

Chicago Historical Bookworks
831 Main St.
Evanston, Il. 60202

Printed in the USA

Illustrations by Scott Holingue

THIS BOOK IS RESPECTFULLY
DEDICATED TO ALL THOSE
WHO QUIETLY PROTECT AND
PRESERVE THE FORESTS AND
THE SEAS AND ALL THE
CREATURES THEREIN.

TABLE OF CONTENTS

Preface

by Robert Cromie

Scott Holingue, a native Chicagoan, ranges the North Pond section of Lincoln Park daily, either alone or accompanied by one or more of his four dogs, all of them leashed, in a search for ailing, wounded, or maltreated birds and animals. The help he offers is immediate and in the past dozen years or so has cost him thousands of dollars. You might call his actions a hobby, but a compulsion is more apt. Holingue, an artist for the Chicago Tribune, cannot bear to ignore a suffering bird or animal.

One day in the early '80s he noticed a bird lying in a doorway on Clark Street west of the park. Its eyes were closed, a wing was broken, and half an inch of bone was protruding. Gangrene already had destroyed several toes. Furious because no one seemed interested, Holingue took the pigeon to the clinic of a noted veterinarian specializing in avian medicine with an office in the north suburbs. After a brief examination the vet suggested euthanasia because of the bird's condition. Holingue insisted that he wanted done whatever could be done, "no matter what it costs."

The bird was named Patton, for clinic records, and after a week of surgery and superb care, Patton and Holingue went home after paying a $260 bill. Patton, obviously unable to compete in the wild, now dwells in a large cage in the Holingue home, next to another cage housing a bird with a broken wing which defied mending.

It was while walking his dogs about 11 p.m. shortly after finding Patton, that Holingue effected his first rescue within the park. They came across a large "an extraordinarily beautiful mallard" huddled at the foot of a tree in the snow. The bird surrendered with only token resistance. The following morning the veterinarian discovered that some "sportsman" had fired a .22 caliber bullet which passed through the mallard's body, just grazing a lung. Holingue treated the bird at home for three weeks, carefully following the veterinarian's instructions on feeding and medication. During the third week the bird was transferred to the bathtub, where it paddled happily, surveyed itself in a mirror, and finally scrambled out. Holingue then took his guest to a southwest suburban nature preserve, where the ex-invalid swam over to other mallards and began courting a pretty female.

"Tales from an Urban Wilderness," written with a Tribune colleague, Kenan Heise, is a combination of anecdote relating to Holingue's continuing efforts to aid the bird and animal residents of Lincoln Park, and descriptions of some of the human frequenters of the same area, including a surprisingly large number of homeless persons. Since Heise not only shares some of his co-author's concern for what happens in the park, but also is one of the truly-knowledgeable writers about the city, the combination is excellent.

The result is a fascinating below-the-surface look at another world, one whose suffering and problems, tragedy and humor, usually are unseen by the thoughtless or the uncaring. Holingue has an on-going campaign against the human and inhumane enemies of the North Pond's ducks, turtles, squirrels, cats, abandoned pets, and a surprising variety of other non-human life. He also resents the city's habit of cutting down too many trees, then turning the wood into sawdust and removing them, instead of permitting fallen trunks and branches to serve as shelter for the area's wildlife. And he was one of the principal reasons why the burning summer of 1988, which turned North Pond into a stagnant and life-threatening morass, wasn't an even worse disaster before the civic authorities replaced the faulty water-flow system and aerated it.

Holingue and Heise, meanwhile, mourn the ignorance of past generations in wiping out such things as the vast bison herds and—in particular—the Passenger Pigeon, once numbered in the billions.

They remind us that the last uncaged member of that species was shot dead in 1900 while the last captive one, named Martha, died in the Cincinnati Zoo in 1914. They also offer a brief history of the century-old Lincoln Park's 59 acres and the 9 1/2 acre segment known as North Pond.

Mention must also be made of a coterie of Holingue's friends, including Mary Jo Nasko, a pushover for any animal in need, who was persuaded to provide overnight shelter for a desperately-ill (and very pregnant) little black cat, and for a splendid Calico cat rescued by Scott as it sat in the middle of a busy intersection. Mary Jo still has both.

Holingue has learned a great deal about the reaction of various wild things, suddenly injured or disoriented. He once saw an injured sea gull sitting quietly on the arm of its rescuer while awaiting treatment at an animal hospital. And most of the sick or wounded victims he finds submit quietly to an examination. Very few of his recovered "patients" show any reluctance to return to their old way of living free. Curiously enough, perhaps the most difficult task of relocating a refugee in a safer area has been provided by a small box turtle that Holingue plucked off a city curb as it was marching at top turtle-speed toward some unknown destination. The little fellow, legs and head safely hidden, soon relished a lettuce diet, but Scott quickly realized that a box turtle has the soul of a Marco Polo and hates to be penned up. Scott's problem-tenant was forever trying to escape, so he took it to the botanical gardens north of the city, only to discover that their ponds, while ideal turtle-havens in one respect, had cement sides which meant turtles couldn't climb out for a needed visit to terra firma. North Pond was too dangerous for turtles because of human tormentors, and a trip to a once-lovely slough in a south suburb found the waters covered with algae and uninhabitable. So at this writing Holingue and his retractable friend still are searching suitable box-turtle territory.

It is difficult to convey the charm of this slim volume, which will reward all readers except those churlish types who lack interest in anything but themselves or one of that "fun-loving" class that isn't happy without something defenseless to shoot at. You will meet a gentle artist who used to grab cabs twice a day to hurry home to minister to a motherless fledgling, who twice in one week had gun-toting teenagers arrested for shooting in the North Pond, and who spent hundreds of dollars on materials from which he fashioned cages for the Trailside Wildlife Museum, a refuge for ill and homeless birds and animals.

There may also be another book in the number of homeless persons who find a refuge of sorts by North Pond. Holingue discovered to his surprise that some of this group were rescuing ill or injured birds and saving them for him, and he was given a name by them. They called him, "The Duck Man."

So, if you are weary of searching through bound-to-bore titles which clutter up the bookstores these days, give *Tales from an Urban Wilderness* a chance. It's a sleeper that will keep you awake.

THE PARK AND POND: AN INTRODUCTION

At North Pond in Chicago's Lincoln Park, wilderness and city meet in a subtle but unrelenting struggle that begins by birth or by chance when a creature first arrives here. You watch as it tries to fit into the world of the humans who, more than a century ago, created the park from a sandy lake shore.

For more than 50 years the North Pond area of Lincoln Park has been my backyard. It is a world of glittering water, craggy old trees and enough vegetation to make this city park a home to thousands of birds and animals as well as insect life. Giant, rough-barked cottonwoods, velvety cat tails and swaying willows add to a feeling of being located in a far northern wilderness rather than in the center of a huge midwestern city. It is the oldest park in Chicago and one that has developed, over the past 100 years, a feeling that somewhere in the early morning mists drifting like smoke around its trees and water plants, lies the soul of a primeval forest. This is a visual paradise of color, light and movement that offers limitless subject matter to

artists, photographers and writers, whose work is often the only way many other people experience nature. But to those who choose to visit the park, the soft silhouettes of geese and ducks wandering through the fog of pre-dawn are not pictures, but real. They are there every day, to be seen free of charge, on Channel Earth.

One morning, I interrupted a conversation with a neighbor to point out the sudden appearance of a Great Blue Heron gliding slowly downward, past the tops of the willow trees that grow along the bank. Watching this giant stork-like bird circling to land on the quiet surface of a pond would be a thrilling sight anywhere, but here, in the city, it especially gives one a feeling not only of excitement, but also of privilege.

Not everyone, however, seems to be aware of this special place.

A bus, filled with yawning workers many of whom wear stereo headphones, accelerates from the Roslyn Place stop on Stockton Drive with a roar that quickly drowns out the sweet songs and cooing of birds going about the daily business of finding food and nesting. Suddenly, nearby trees, as if in defense of the tranquility of the park, shake their leaves, driven by strong lake-bound winds, in a hissing crescendo that seems to scold the bus for its intrusion.

Many trees and plants now grow wild here, but, for the most part, Lincoln Park is a man-made garden. Built by hand-labor upon the route of an ancient north-south flyway that only a few generations ago witnessed the migration of practically limitless flocks of birds, the park area today still offers its water and trees as a refuge and rest stop. A much-reduced population of birds passes through each Spring and Fall with most continuing on to seasonal destinations. But some decide to stay for the summer and raise their young . The odds for survival are never in their favor. Aware of their struggle, the author has chosen to become an occasional caregiver to feathered travelers and to some of the creatures who are permanent park residents.

The Park and Pond: An Introduction

Living next to this part of the park since I was born more than 50 years ago has not lessened the sense of awe I first felt as a child while wandering beneath it's huge cabbage-like trees, observing up close nature's ways of survival and growth. The creatures I have seen and fed there each day seemed to observe me with much the same sense of hesitant wonder as I have them.

The 59 acres of Lincoln Park that lie between Diversey Parkway on the north, Fullerton Avenue on the south and Lake View Avenue on the west are part of the city's front lawn on Lake Michigan, one of the largest bodies of fresh water in the world. This is a very complex park, a hybrid of city and nature, of metropolis and lake.

As such, the North Pond area is a storybook of nature, a laboratory of life, but most of all an unfenced home for both animals and humans.

The pond area is a step-sister of Lincoln Park's more formal southern section, which contains a conservatory, zoo, boat pond, farm and athletic fields. The 9 1/2 acre North Pond, for being part of the city, offers a surprisingly wild appearance. Tall, billowing trees along with untrimmed bushes and flowering plants line most of its banks. A steep, unkempt shoreline slopes upward to a 20-foot high ridge on the east and recalls that 150 years ago the area was a rolling landscape of shifting sand dunes. As precious and beautiful as natural seashores are to us today, it was then felt that a growing city needed something "more civilized" for the use of its citizens.

Using horse-drawn wagons, workers laboriously added naturally fertilized topsoil, trees and shrubs to a dry sand surface that for thousands of years had supported a dune ecology of wild grasses, bristly plants and a few scrub oaks.

The park, originally conceived in the mid-nineteenth century as a recreational setting for Chicago's citizens, has evolved into a year-round habitat for birds and animals who often find themselves competing with humans for a place in the natural tranquility that exists there. My love for this area and its creatures includes a certain possessiveness: a distinct, protective watchfulness, developed over a lifetime of observing its fragile life cycles.

To the creature fortunate enough to survive to adulthood, remaining a year-around park resident can be difficult. Decreasing numbers of old hollow trees in which to hibernate or just rest in winter often leaves hungry squirrels and birds out in a sometimes 20-degree below zero cold. Failing to find food beneath ice-hardened snow, they become dependent upon the generosity of those people who leave them bread and seeds. Sadly, the spring thaw each year reveals that some found the struggle too difficult to overcome.

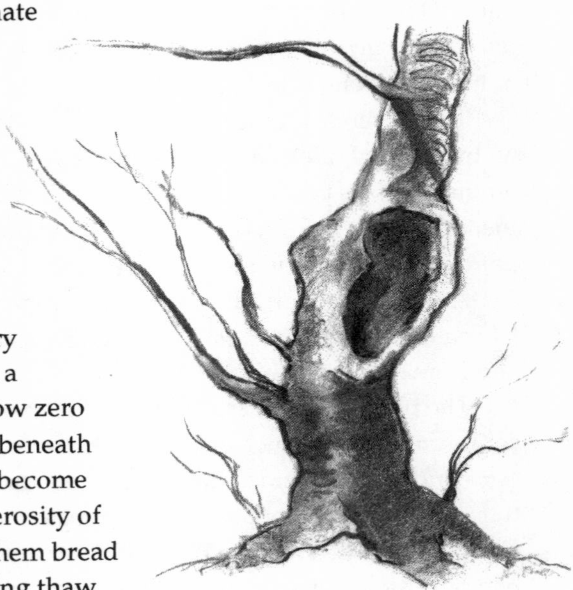

In the opposite, but equally difficult, weather extreme, young ducks and geese growing up in 100 degree summer heat search continually to find edible plants in what can be stagnant, polluted pond water.

Ironically, extremes of weather are welcomed by many of those concerned with the increasing overuse of the park. To them, it seems, the cold, rain and wind—anything that will mean less human contact, can let this forest-like garden and its pond continue to develop its true wildness.

Here at North Pond in Chicago's Lincoln Park, wilderness and city meet in a subtle but unrelenting struggle that begins by birth or by chance when a creature first arrives here. You watch as it tries to fit into the world of the humans who, more than a century ago, created the park from a sandy lake shore.

I learn from each experience, whether propping up an old tree or watching a veterinarian set a broken bone in a wing. It drives me to help retain the balance in favor of the park's natural environment and of the wild creatures who make it their home.

The conflict often is one of human caprice and carelessness against the forces of nature. But it is not simply a story of humans always being the bad guys. Many people do help. They offer food to the ducks, squirrels, songbirds, and the various species of wild doves. They pick up litter. They call the park district if something is awry. And some, not knowing what else to do, notify me when they see a bird or animal hurt. Mainly, most of them just give the animals their space and there is much to be said for doing just that.

Others do not. They see the park and its pond as an opportunity for abuse rather than use. They clutter. They adopt a sadistic attitude toward the wildlife they see. Sticks and stones do break bones when the victim weighs but a few ounces. Perhaps the attitudes of some of these people would be different if the park were in a formally designated wilderness area. Just possibly, they might be more in awe if that were so. But, as it is, the park is part of the city. It is not only used by taxpayers and the public, but also owned by them. They see this as their property and feel they have a right to do anything they choose here.

Such individuals see the wildlife, especially the grey squirrels and European rock doves, commonly called pigeons, as intruders, as having no rights, as being little more alive with the spark of creation than the metal ducks, birds and animals in a shooting gallery.

We as a people have drifted a long way from the beliefs of those Native Americans who had resided in this area since the age of the glaciers and until a comparatively short time ago lived on the very ground that has become this city.

They had a bond with nature and with others animals who shared their space, even those which they had to hunt for food or clothing. To these, they apologized when they needed to take their lives. They named their clans after the wild creatures and saw themselves as one with and descended from them.

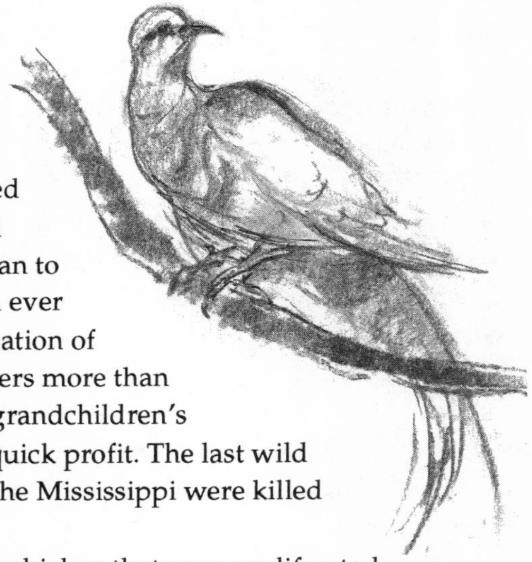

Defenseless against human beings, many species from the buffalo to the mink that once flourished in the Midwest disappeared from the area after they began to be extensively hunted by an ever westward-expanding population of settlers, soldiers and profiteers more than willing to trade their great-grandchildren's pristine environment for a quick profit. The last wild buffalo and her calf east of the Mississippi were killed in 1825.

Birds such as the prairie chicken that once proliferated here quickly became extinct in the area, as did the long-tailed graceful pigeon that the Eastern Narragansett Indians knew as "Wuskowhan," the wanderer, and observed swirling majestically in flocks so large that they darkened the sky. That bird was the iridescent, long-feathered Passenger Pigeon, whose kind had lived unmolested for 2 million years. It was reduced, gruesomely, by market hunters, and those who simply wanted them destroyed as a nuisance. They went from a wild population numbering in the billions to just two birds, one living in the wild and one in captivity. Officially, the last member of its species, flying free, was shot dead in

Spring, 1900. That left just one, a female named "Martha" by her captors, who survived until 1914 when she died at the age of 29 in a Cincinnati zoo cage.

One story of the North Pond is that somehow, every once in a while, a species—one that long ago abandoned the North Pond area as it became "citified"—returns. Tales will be told in these pages of such creatures being seen here, including a beaver who simply appeared one day and stayed nearly a year, woodpeckers, cranes, seldom-seen species of songbirds and even the previously mentioned Great Blue Herons, who mingle seemingly with little fear among the people who stop to marvel at their presence.

The struggle for them is to try to maintain a continuity of existence from ages long ago through into a future that is very uncertain.

This place was set aside by people long before they could possibly have known just how urgent environmental issues would become. We have a distance to go before our actions, as well as our words, acknowledge that urgency. It is not enough to campaign to preserve the national parks and their forests and wildlife or to save the wetlands and state preservation areas; we need do the same for the life, both plant and animal, in our city parks. The harmony among their trees, shrubs and water calls for our respect as well as enjoyment. Bonds with the natural order that Native Americans knew and early

settlers acknowledged were not imagined, but real.

The future of this neighborhood and of our city is bound together with this half-square mile piece of land. The story of the park is that life is bounteous, beautiful and good and also interdependent—a message that is heard from the pond and its wildlife.

MOTHER BIRD

The bird remained quiet under my desk
and I assumed it would remain so.
How wrong I was!

Feeling that each life saved is important, I decided that the next time I saw a baby bird obviously abandoned I would attempt to save it. The following story tells of just such a rescue and contains a somewhat embarrassing moment that happened as I tried not only to help parent an orphaned nestling but also have that effort fit in with the routine of my daily life. The incident illustrates both the seriousness and the humor that can so often arise when we humans try to help other creatures.

While a mother bird will feed her offspring as frequently as every few minutes, I have found that the best I could do for a baby grackle, rescued one afternoon from the path of the park mowers, was to feed it once each hour, my schedule permitting. Being just a part-time provider proved to be difficult and awkward. I guess, however, that one runs such risks when trying to be the parent of anything.

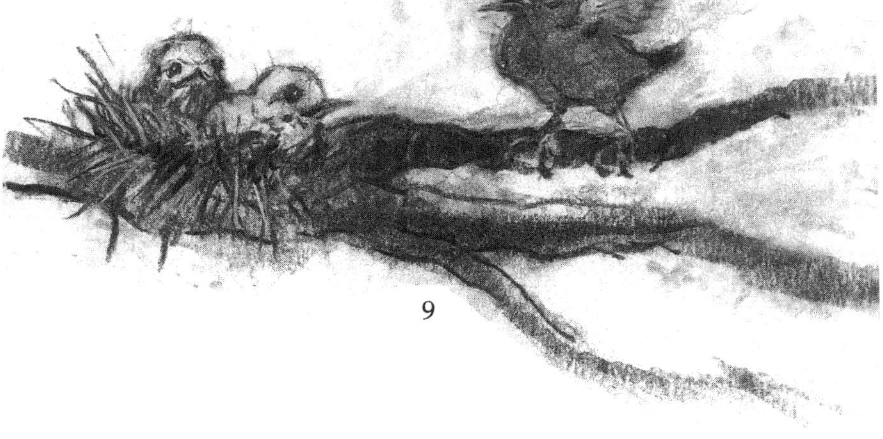

One Saturday afternoon the fledgling undoubtedly had been exploring the world around its nest located high in a craggy "old-growth" tree and had fallen to the ground. Saved from injury by instinctively flapping its tiny wings it had landed on a soft cushion of uncut grass. By the time I arrived the bird had clearly been abandoned by a mother probably too terrified to fly to the ground to feed one of her babies in what was a very busy section of the park. Most certainly there were hungry brothers and sisters to care for still in the nest. The frightened little grackle, its wings not yet feathered enough to allow flight, tried to hop away as I approached. Since the uncut grass was higher that its tiny head, it proved easy to catch. In my hand I could feel its tiny heart pounding. In the wild, predators capture their prey in much the same way. This time however, a little bird—normally doomed under such circumstances—was being given a second chance to survive.

In my home when I first started feeding the frightened youngster, I tried to do so at least every half hour. I arose at dawn and continued until after dark, trying to maintain the rhythms to which it was accustomed outside. As the weekend passed I quickly realized that I would soon have a new problem to solve: how to feed my guest and also work at a job three miles from my home. At first, during the early part of the week, I rushed back and forth in cabs several times a day. This proved to be both costly and inconvenient. One solution I tried was to ask for help from my friend, Mary Jo. She agreed to do what she could.

I would feed the bird from early morning through noon and she, working an earlier shift than I did, would take over late afternoon through early evening. It didn't always work out as planned. One day she was unable to find a bus that wasn't packed to the doors with people or an available cab. Being very concerned for the welfare of that little bird, she wound up walking three miles to my home to tend to the very hungry boarder. There had to be a more convenient way and I decided that the best solution was to bring the bird to work with me.

A mother bird feeds her little ones by pushing food far down their throats in a manner that at first appears to be rather rough. It really isn't. It's just her way of ensuring that the food gets down to a place where it is easy to swallow. She will continue to give them food until they appear to be well into adulthood. In her place, I had to do the same thing, or at least the next best thing I could contrive. I chose at first to use a powdered nutrient made specifically for raising baby birds and available in pet supply stores. When mixed with water and heated for a few minutes it takes on the consistency of thick soup. This mixture of proteins, vitamins and other nutrients is necessary for young birds to survive the first few weeks of life. It also has proven to be an effective emergency food for injured or sick adult birds too weak to feed themselves. It offers them nourishment until they can eat normally again. Using a plastic eye-dropper I was able to administer the food one drop at a time, pausing to allow the bird to swallow each one before continuing. After two weeks or so of being hand-fed, a bird typically begins to show signs of wanting to eat by itself. Then more solid food can be provided. Later on, it will eat grubs and other protein that would be found in the wild. At that point, a ration of edibles placed in the cage each morning will last until night time.

This particular experience, however, was one of my first and I was unsure how to transport and feed a baby bird while working in a business office side by side with many other people. I did not want them to know what I was doing. The solution I decided upon was to modify a 3-quart plastic food container by drilling 1/2 inch air-holes in its removable lid and placing several folded paper towels as a cushion on the bottom. It

worked well and, once the bird was inside the container, it usually remained quiet. A small nylon backpack with room inside for the container and the feeding kit proved to be an ideal means of carrying my puzzled pal to work with me.

The next morning I boarded the bus to work and found a seat. I carefully balanced the special bundle on my lap, trying to keep the usual shakes and bounces from traumatizing my patient. Arriving at work, I decided to keep the backpack on the floor beneath my drawing board until feeding time. To perform the actual feeding I carried the pack each hour into the washroom. I felt that, in corporate America, it would just not do to be observed sitting at my work station prying a baby bird's beak open and stuffing food down its throat. So, I took the whole production, backpack, bird, food and all into one of the men's room stalls, closed the door, sat down and proceeded with the feeding.

This generally worked well, although some people must have thought it odd that I was going to the restroom so often, and carrying a backpack yet! The bird remained quiet while under my desk, and I assumed that this perfect behavior would continue throughout the day's feeding. How wrong I was!

That afternoon, as I held the little bird and prepared to provide its meal, someone entered the restroom.

A bit nervous, I fumbled with the eyedropper and took a few seconds longer than usual to fill it. A tense moment arose when the hungry little bird, impatient at not getting its food right away, decided to complain as it would when in the nest. As I filled the eyedropper from the small bottle of nutrient, the grackle let out a loud, screeching squawk.

At that moment I sincerely wished that I was somewhere else very far away. Through a crack in the door I could see the person who had been washing his hands pause and slowly turn his head in my direction. His expression was unforgettable.

Back in the stall, I looked down at the bird with eyes that said, "Please..... don't .....do.....that.....again."

The little grackle looked right back at me, closed its gaping mouth and mercifully chose not to squawk again. After barely pausing to dry his hands the man hurriedly left. I finished the feeding, packed up and returned to my desk. Sitting down, I mused that if fate had twisted the other way and the puzzled gentleman had waited to see who or what emerged from the stall, I was fully prepared to walk out clutching my bag and repeating the noise myself as I went about clearing my throat.

After maturing, this bird was successfully released from my backyard and flew straight to the park to be with its own kind.

THE NORTH POND

In 1874, it was reported, 37 pairs of English sparrows were purchased at $1.50 per pair from New York City and released in the park.

It is hard to imagine the North Pond area of Lincoln Park as having once been outside Chicago's city limits, but in 1852 it was. That year, a cholera epidemic reached its peak and swept unchallenged throughout the city. Unlike today, there were no miracle drugs and very few effective supportive measures to keep people from dying. It was concluded that by isolating the sick perhaps the disease could be brought under control. A panicked citizenry demanded that something be done, and in response, Chicago's Common Council decided to purchase three large tracts of land outside the city to be used for the construction of hospitals and quarantine stations. One of these parcels consisted of 59 acres of sand and marsh that supported only a few bushes and scrub oaks and was intersected by a long, stagnant drainage canal. For this unappealing section of lakefront, the city paid $8,851.50, a sum that today would hardly pay the real estate taxes on a medium-priced home located adjacent to the park. Ironically, the epidemic subsided soon afterward and the

site was never used for the purpose for which it was purchased.

As early as 1865 the city had made surveys to improve the "old park," as it was called. The land remained virtually unused for recreation until 1869 when the Council voted to create three park districts on the north, south and west sides of the city. The "Lincoln Park Act" was the first to pass, being approved on February 8, 1869, and included the land north of Fullerton and east of Lake View Avenue. Other than location, the area was a landscaper's nightmare because of constantly shifting sand dunes upon which only a few species of wild grass and plants would grow. A plan for development was not passed until 1881, when a number of prominent Lake View citizens petitioned the city to make improvements that included a "lake," as well as "lawns."

Although a picnic area was created in 1871 after many forest trees were planted between Stockton Drive and Lake View Avenue, the stagnant water in the drainage canal known as "the 10-mile ditch" reminded residents that the area was still in part a low-lying marsh. Finally, after receiving many complaints, the commissioners undertook the filling of the canal and, in 1872, began the construction of the the North Pond. Much of the sand excavated from the pond project was sold but enough was kept to construct a large hill to the north and this was called, "Mount Prospect." Today, a statue of turn-of-the-century Illinois gov. Richard J. Ogleby facing Lake Michigan stands on top of this mound, known in my childhood in the 1950s as "Devil's Hill."

Rich black soil from the smelly drainage ditch was used for the construction of grassy lawns on either side of the pond. In exchange, the sand from the pond was used to fill in the ditch. In 1884 the North Pond was completed and the water valve at the north end was turned on to fill the 9 1/2 acre excavation. A two-story high pavilion was built at the north bank and used as a boat house. In June of that year row boats were moored in a circular basin and soon people were rowing across the pond's smooth surface. Drives were built and carriages were for a time able to ascend the heights of Mount Prospect for a panoramic view of Lake Michigan. Buggy rides had to be halted when rainstorms eventually created huge gullies alongside the carriage paths causing them to be abandoned and sodded over. Along Lake View Avenue, long-standing stables, shops and sheds were torn down in 1889 when the street was officially opened as the western border of the park.

To create a soil base for turning the park into a garden, 150,000 cubic yards of clay and black soil along with thousands of wagonloads of manure were spread upon the surface of the sand in hopes that trees, grass and flowers would take root and grow. The plan worked so well that 200 "Keep off the grass" signs were put up, interestingly half of them in English and half in German to accommodate the North Side's large German population. Some 20,000 trees and 100,000 plants were purchased to ornament the park replacing hardy scrub oak trees that were cut down by the city.

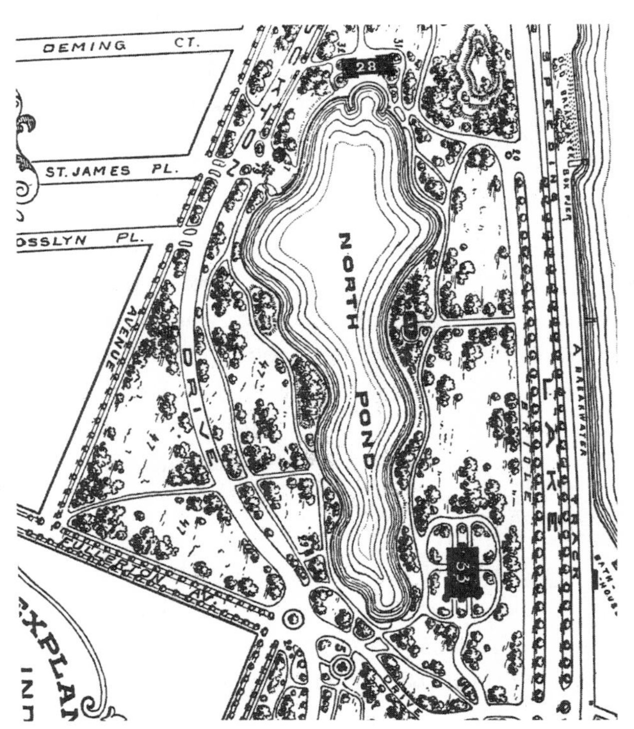

Near the northwestern bank of the pond an artesian well 1,540 feet deep was created that daily attracted hundreds of people who filled their jugs with its pure water. The above-ground portion of the well was constructed of stone and resembled a miniature mountain ten-feet high by ten-feet wide. Fame proved to be its undoing, however, and after only six years of operation, the water flow, once an impressive 332,352 gallons per day, had diminished to the point that the commissioners approved a plan for the laying of city water mains to serve the park's increasing needs not only for drinking water but also for sprinkling. Today, only a small patch of cement marks the spot where once pure, clean water was enjoyed by thousands of park visitors.

Goldfish and carp were introduced to the North Pond and concern for their welfare was such that permission was denied to fly-cast on the grounds that it was considered "an insidious attempt on the peace and happiness" of the goldfish in the park pond.

In 1874 at the cost of $1.50 per pair, 74 English Sparrows were purchased from New York City and released in the park.

A park police force was created and it is noteworthy that in the late 1880s they were bicycle-mounted. Chicago police in the park today are also patrolling on bicycles as part of a "new" scheme to reduce crime. The service was created in 1872 and among its various duties was "to protect flowers, birds and squirrels from the assaults of small boys."

Ironically, one of the laws passed in 1882 by the Council banned the use of bicycles in the park by private citizens after a high-wheeled bicycle rider caused a horse to panic and become a runaway. This ban, with a few exceptions for celebrations, lasted until 1886 when "wheelmen" pointed out that horses were now more "civilized" and used to the sight of bicycles.

As the century rolled over, the park continued to be improved by the addition of first class gardens that included a world famous half-acre lily pond visited by admirers from as far away as Japan.

Lake Michigan's beaches and breezes were not only an attraction for those seeking recreation, but, for some, a necessity for life itself. A large screened pavilion, still standing, was built on Lake Michigan at Fullerton Avenue in 1889 and used as a fresh-air sanitarium for the benefit of sick babies and their mothers seeking a place to recover outside of the sweltering heat and pollution of the city.

The early part of this century marked the beginning of a decline in the care and pampering of the park with many of the smaller gardens falling into decay. The era of daily weeding of the park by dedicated gardeners was over. The landscaped garden walkways in front of the conservatory became overgrown and abandoned. Its boulders and rocks soon became lost in piles of leaves and trash thrown into the bushes by insensitive park visitors.

During hot July nights in the 1920s my grandparents and their neighbors thought nothing of bringing a blanket to the park and sleeping outside to find relief from Chicago's infamous heat and humidity. Today, only the homeless stay overnight in the park.

During World War II, the sandy soil was still rich enough to grow vegetables in a Victory Garden, planted in the wedge-shaped area between Diversey and Wrightwood, Lake View and Stockton.

The North Pond

In the 1950s, the Park District police force was disbanded and responsibility for patrolling the park fell to the Chicago city police. The old park police station still stands east of Clark Street at Armitage Avenue and is now a community center.

PATTON

I told him I did not want the bird to die.

Giving human names to birds seems to demean them. I don't do it. But, with the initial two birds that I took in, veterinary records required a name, so I quickly suggested "Patton," after the tough World War II general, because of all the war and ravage this pigeon had survived. The second, whose story I will tell later, was called "Grumpy" by Mary Jo's mother, who bird-sat for us during one rare vacation. It does describe his normal demeanor.

It was twelve years ago and as I walked passed a doorway on Clark Street just west of the park, I noticed a bird lying on its side with its eyes closed. A wing had been broken and a half-inch section of dried bone was sticking straight out, indicating that the injury had occurred at least a few days earlier. Pathetically, his feet were gangrenous and several toes had already rotted off.

I felt a sense of outrage at the apparent non-concern of passersby and became determined that this bird was not going to die in a doorway on Clark Street. I would do what I had to do to save it. I took the pigeon to a nationally-known bird veterinarian, a specialist who operated at that time in a suburb north of Chicago. At the clinic I placed the bird on the examining table and sat down. The veterinarian's face seemed to sag a little. At first his recommendation was euthanasia, saying that the bird was in such poor condition. There was no way I would agree to that without having to at least tried to save its life. I told him I did not want the bird to die.

"I don't care what it costs," I said.

Scrutinizing the battered creature's injuries, he sort of hummed to himself, then said, "OK," and handed it over to a young assistant to care for.

And care he did.

It cost me $260 for a week's worth of hospital care and surgeries, a lot of money, but Patton is still alive and well today, these many years later. And I have enjoyed his companionship, although he still greets me with a growl and a slap of the wing.

The bird is a pigeon and that is what he is. I sometimes hesitate to use the word "pigeon," because that name has become disparaging and pejorative. It is as though to many people they are not real birds, not individuals. The problem is not with this bird, but with the public which has created contempt out of closeness and familiarity. They are individuals, as birds, as doves—which they are—and as pigeons.

Many people see these birds simply as nuisances that proliferate with absolute abandon rather than as individual creatures, who, like other wild birds, have only a 30 percent chance of surviving their first year of life.

They are an immigrant bird descended from an ancient species of dove that has existed for 20 thousand centuries, choosing to nest in the rocky cliffs of Europe at a time when humans were still living in caves. They have personalities and coloring that vary greatly among individuals. Theirs is not an easy life. Like other city birds, they can suffer frostbite, for example, and have their toes very painfully frozen off. They can get stuck in commercial repellents placed upon buildings. There, unable to escape or grasp the reasons for it, they slowly die of thirst.

Patton

In the world of nature, appearance is everything. Birds that are hurt or injured can be easy prey to hungry animals and even other birds. Out of dire necessity, they will use up every last bit of energy to maintain a front of normalcy, not wishing to attract attention by looking vulnerable. Sometimes they recover from an illness or injury, but other times they do not. An exhausted sparrow once flew up to me, drew its wings to its side, then dropped to the sidewalk. At first, it seemed to have died for no reason. But as I turned it over, I saw that it had an old badly infected eye injury that had weakened the bird to the point where it had no more energy reserves to maintain activity. It used up its last moments of life to make one final flight, probably sensing somehow that death was imminent and believing that in the air there was safety .

With Patton, it was my pleasure to see him go from such a state of utter desperation to the warmth of an incubator in a veterinarian's hospital room with his wounds bound up and hunger being satisfied with healthy food. It was like his waking up in a MASH unit after a battle.

Healthy, alert and proud looking with his now puffed-up chest, Patton became a permanent resident in my home where he is living out his days. He has a large, three-foot square cage, with tree limbs and a miniature pond made from a wash basin. He could not be released because his shattered wing, missing a key section of bone, would never function normally and support his weight in flight. A solitary bird in a cage to me is a sad sight so I was pleased to find what I thought would be a mate for him. But companionship for Patton was something else. The only other bird I ever named—which, as I said, I don't like to do—was the second pigeon I took in. He also had a broken wing that was irreparable. I accepted his christening of "Grumpy." He always seems to have a chip on his shoulder. A pigeon, wishing to appear fearsome, can growl like a dog, giving off a guttural sound. Whenever I come into the room to feed them, Grumpy starts to tremble and, if I do not get too close, goes into an aggressive little strut, like a war dance, trying to scare me away. He turns around and around and makes threatening noises.

At first, thinking he was a female, I put Grumpy with Patton in the same cage. It was a mistake. They were too much alike. They immediately charged at each other, locked beaks twisting and writhing on the cage bottom until I pulled them apart. Finally, I built a second large cage. They are adjacent to each other, but cannot get at one another, so they do have companionship of sorts.

My First Patient From the Park

It seems among wilderness-born creatures that once health is restored, the need to be free is so great that captivity creates an intolerable stress. It was time to set him free.

The first time I saw an injured duck, I wasn't, at first, able to tell quite what it was. It happened one winter night about a dozen years ago and had the effect of altering forever my relationship with wildlife. Caring for that wonderful bird gave me the confidence to try to make a difference to any creature I found that was in need.

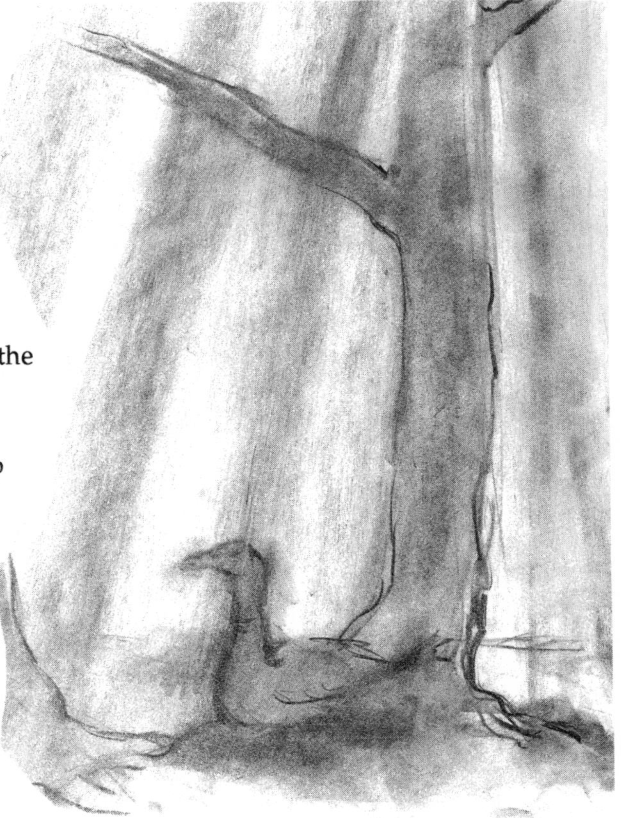

At night in the park, sodium-vapor street lights cast a yellowish-white light so intense that it is sometimes mistaken for daylight by birds. Where it shines through the branches of trees, the effect on the ground is one of a jig-saw pattern of bright light and deep bluish-grey shadows. It is easy to overlook anything hiding there.

About 11 p.m., I was walking with my dogs on the sidewalk between Stockton Drive and the North Pond when we noticed movement in the snow near the the base of a tree that stood next to the curb. We stepped off the sidewalk and, apprehensively, took a few steps closer to see what it was. A sound of coughing and wheezing came from the shadows. My dogs raised their ears and stared at the changing shapes in the snow. Just as I leaned forward, a large, extraordinarily beautiful mallard shuffled out into an area of bright light. Untypically colored, with feather hues more like those of a Canada goose, he presented an impressive sight. His head was black and on the front part of the neck was a white band, instead of the iridescent green heads and multicolored bodies one usually sees on a male mallard. The rest of the feathers were greyish-tan. In one of my books on waterfowl I once read that mallards can mate with other duck species and produce striking color variations. But with a Canada goose?

As I moved closer, he raised his head high and struggled to stand. Instead of jumping straight up and flying off as a normal duck would have done when threatened, he tried to escape by running through the deep snow. Laboring to breath he was unable to take more than a few steps before stumbling and sitting back down. At first, judging by the sounds he made, it seemed that the duck had been weakened by a respiratory infection, possibly pneumonia, or some other sickness.

With no previous experience with injured ducks, I nevertheless strongly felt that I wanted to help this one, whose plight seemed so desperate. Free-running dogs, or other predators, would kill the duck if I didn't do something immediately. My own dogs still watched with restrained curiosity but their very presence was alarming to the mallard. I took them home, where I searched for a pair of gloves and a bag.

When I returned, the duck was still there sitting in the snow. He again struggled to get away but was too weak to go more than a few feet. When the frightened bird tripped and fell in the snow, I quickly reached down and picked him up. Able to make only weak quacking sounds, he barely struggled as I placed one hand under his body and restrained his wings with the other. Holding his head high and watching my eyes during the walk home, he appeared frightened, but seemed to sense that my actions were not those of a predator. I once saw a seagull with a broken wing that remained perfectly calm while standing on the arm of its rescuer in a veterinarian's waiting room. That man told me that he had experienced similar reactions from other injured seabirds that he had saved near his home on Lake Michigan.

All the way home the duck never stopped staring at me. In the front hall of my house, I had to squirm past my curious dogs and carry him upstairs to a spare room where I could try and figure out what was wrong. I carefully placed the confused duck on a towel and looked him over, paying special attention to any discoloration that might indicate a wound. Sometimes it is very difficult to see an injury owing to the very dense layer of insulating feathers that covers a duck's body. They must be pushed aside with the fingertips in order to see the skin. It was while doing this that I noticed a single drop of blood on the underside of one wing. Any sign of blood is always a cause for concern, yet further examination revealed no obvious wound or broken bones. Not wishing to cause the mallard further stress, I chose not to place him on his back although that would have allowed me to check the more hidden areas of the wings and stomach. At 7 o'clock the next morning, I put the mallard in a box and drove him to see the same veterinarian who helped me with the pigeon. He was one of only a very few avian specialists in Illinois and had been practicing for many years.

The veterinarian put the box on the examining table and opened it. In the bright light of the examining room, we were both struck by the extraordinary beauty of this bird as he assumed a head-high position while watching us. He was at least alert, the veterinarian noted, and that was a very good sign.

He carefully lifted the bird out of the box, set him on the stainless steel table top and began to examine every feather from front to back looking for a reason why this creature was ill and why there was a sign of blood on that one wing. The anxious mallard breathed rapidly, still making the type of gurgling sounds that accompany lung congestion. He remained still, however, and let the veterinarian check his head, eyes and throat without struggling. Finding nothing obviously wrong there, the doctor lifted the duck, placed it upside down on a towel and examined its underside and looked at its wings.

After a few seconds, the veterinarian grimaced and said, "Here's the trouble."

Motioning for me to lean forward, the doctor pointed to two pencil-sized holes, one on each side of the bird, under its wings—where a human would have armpits. A .22 caliber bullet had passed clear through the mallard, nipping his lungs, but miraculously not piercing any other vital organs. Someone, somewhere, shot him in flight, yet he managed to continue on, landing in Lincoln Park when he could go no farther. It was ironic that his strength gave out just at a place where he would cross paths with someone who would help him. What sense did he make of the strange behavior of these humans? One attempts to kill him, and another tries to save his life. He must have found this experience very confusing.

After cleaning the wound by removing

bits of broken feathers and dirt that had been forced inside, a veterinary technician showed me how to feed the duck using special feeding tubes and syringes. Confident that I could do the job, the veterinarian released the bird to my care and I took him home. During three weeks of treatment, with continued veterinary supervision, I watched the mallard gain strength and fully recover.

At first, to prevent him from aggravating his wounds by trying to fly, I placed him in a cardboard box just large enough for him to stand and turn around in. After several days, when it appeared that he was comfortable with captivity, I replaced the small box with a slightly larger one. He slept on thick, folded bath towels and, several times a day, was served breakfast in bed—my bed.

As his strength returned, giving medication and tube feeding him became a real battle. I had to pick him up carefully but firmly , hold him under my left arm like a football and try to avoid being scratched by his claws or pinched by his now-powerful beak. Extending his neck to its full length, I held his head with my left hand, gently pried open his beak, and very carefully inserted a soft rubber feeding tube connected to a large syringe full of nutrient down his throat into the crop, a stomach-like pouch used for storing food. Extreme care had to be taken not to overfeed or to place the feeding tube into the breathing passage which would have caused the bird to choke and possibly die.

The daily cleaning of the bullet holes included squirting an antibiotic ointment inside to prevent infection. His recovery depended upon me doing this a minimum of three times each day.

And recover he did. One day he began eating the food pellets and fresh vegetables that I provided each day in anticipation of his regaining the ability to eat unassisted. At the end of two weeks, his bullet wounds had completely healed. No longer making wheezing sounds, his lungs appeared completely recovered. With his wounds now closed, it was time to allow him to swim in a bathtub partially filled with water. After placing him in the tub, I turned on the water. The noise of the streaming, bubbling faucet startled him. He watched as the water slowly moved towards his toes. When it reached his feet and then legs, he began rapidly wagging his tail feathers, seemingly in anticipation of having some fun. Slowly, the water level rose enough to lift him upward. His toes began scraping the bottom of the tub and soon he was able to move under his own power. Within the confines of the tub, he swam forward, backward and in tight circles, pausing to dunk his head and exhale air underwater making happy bubbly sounds. Suddenly, he stood straight up, noisily flapped his wings and quacked for joy. He was almost home! It was thrilling to watch him dive forward, submerge his head and whip it from side to side churning the tub water into a miniature storm. After each session I would drain the tub and replace his food and water containers.

My new morning routine consisted of getting up, entering the bathroom, lifting the duck out of the tub, placing him on the floor, cleaning the tub, then showering amidst increasingly louder complaining quacks. I wondered if my neighbors were puzzled by the strange sounds coming from my second story bathroom window.

Thinking that it would provide him with a form of company, I placed a large hand mirror upright on the bathroom floor next to the wall where I kept his food bowl. It soon became the focal point of his post-swim preening. When he grew tired of swimming, he would stretch his neck up until he could see the floor beyond the rim of the bathtub, then, after rocking back and forth a few times, as through counting, "one...two...three!," he would suddenly half-fly, half-jump up over the tub rim onto the floor. Sitting down in front of the mirror, he would squeeze excess water from each feather with his beak, all the while chatting with his own image probably wondering why his companion was so agreeable. This went on for about a week.

Then, one morning as I entered the bathroom, instead of standing up and greeting me with quacks and chortles, he just sat quietly next to the mirror and took only slight notice of my presence. He made soft guttural sounds.

Something was wrong, and I felt that I knew what it was. I'd gotten to know my little friend pretty well over the past three weeks and my instincts told me that, although he had recovered from his wounds, this citizen of the wilderness was now on the flip side of captivity. It was the side that causes wild things, needing to be free, to become dangerously depressed when kept in confinement. Although he had regained nearly all of his strength and was eating on his own, he seemed to become more quiet with each passing day and to be losing the enthusiasm displayed during healing. I became alarmed. It seems among wilderness-born creatures that once health is restored, the need to be free is so great that captivity creates an intolerable stress. It was time to set him free.

Confident that the bird's strength and health were perfectly restored, I placed him in a cardboard box and drove to a nature preserve in the southwestern suburbs where many waterfowl, including mallards, lived. It seemed a perfect place to release him, being out of the city, and located along the ancient flyway south.

At the slough, a large body of water containing an abundance of edible water plants, I carried the box to the water's edge and removed him. Holding him for a second or two, I looked at his now very bright and alert eyes and felt a deep sadness, not only at having to say goodbye to this wonderful being whose recovery had given me so much pleasure, but also because I wondered if he would survive at all. He had paid his dues, and deserved to live out the remainder of his life unmolested.

I knelt down and extended my arms. When I released my hold, he vigorously flapped his strong wings and jumped to the ground. He looked as though he could not believe his good fortune. Wind!

Water! Ducks! Where was the bathtub? Those ducks over there surely weren't in any mirror! He walked into the closest water and swam towards the plant cover growing near the bank as though needing a place to assess exactly what had happened. Nearby, other mallards took notice and soon one of them swam over as though he were an emissary sent to check out the new guy. A conversation ensued of soft quacking. The resident duck appeared to be explaining the rules of the slough to the newest visitor. My former patient then swam out away from the reeds near the bank to clear water and in a burst of pure joy began a speedy, flapping run along the surface fully testing his wings for the first time in three weeks. He then swam back to a group of other mallards where he noticed a speckled female swimming nearby. In an explosion of water, he proceeded to chase her, nipping at her tailfeathers.

I had a feeling that with his priorities now in place, he would indeed have a chance to survive. That was the last time I saw him.

Each time I returned to visit in hopes of spotting him, I first stopped at a small grocery store in a nearby town and bought twenty heads of lettuce to feed the waterfowl at the slough. I returned time and again to check up on the mallard, but in spite of his unique coloring, I could never be entirely sure of his identity. There were many, many ducks there.

During one visit I saw a solitary one out in the middle of the slough, which had the same markings and which, I believed, could have been my former house guest.

The following summer, there was a terrible drought that severely affected the wildlife at the slough. The water level was much lower than before and the thick, green algae that formed on the surface did not allow sunlight to penetrate the water to the plant growth on the bottom. Soon the aquatic vegetation disappeared and the fish and turtles that depended upon these plants for sustenance were in great trouble. A few concerned people would come by each day with lettuce and bread for the remaining few ducks who tried to live there.

One day I saw one of the most unusual sights I have ever witnessed at a body of water. As the donated bread was being tossed to the ducks standing on the shore, dozens of fish wiggled up onto the bank until they were able to reach the bread thrown to the ducks! They then made there way back to the water.

Because that year the number of mallards was greatly reduced, I like to think that I helped their population by giving at least one of them another summer to live.

For me, it was a positive experience, one that I remember fondly. It was the first of many, ultimately rewarding, encounters I would have with park life and from that time I resolved never to ignore a creature in distress.

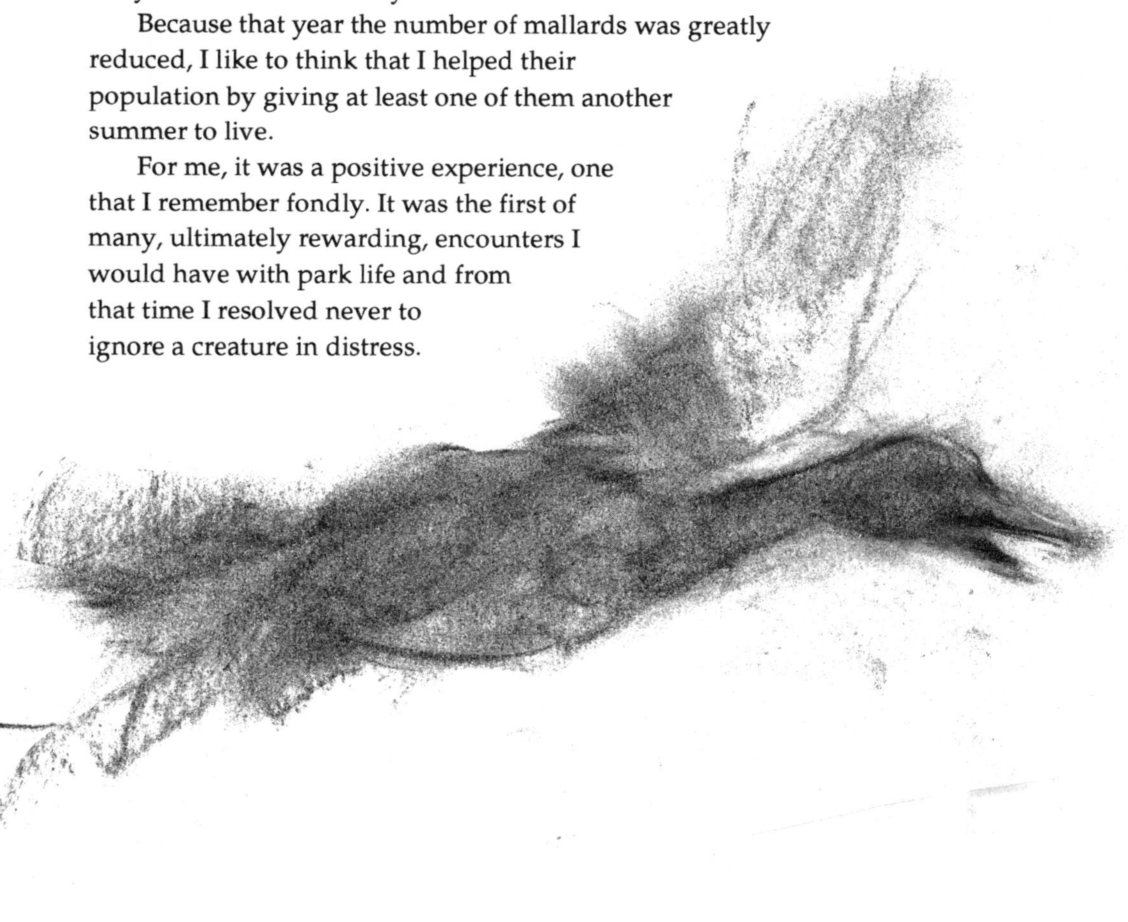

THE DUCK MAN

It was an honorable title and I felt flattered to be recognized unofficially as patron of the park's waterfowl.

The drought that year that affected the North Pond caused me to become known to the homeless in the park as "The Duck Man." It was an honorable title and I felt flattered to be unofficially recognized as the patron of the park's waterfowl. These particular people, at least, seemed to have a sense of charity toward the wildlife there, a willingness to share with them what little they had. Among some of the homeless you sometimes see not only a certain camaraderie among them, but also a sympathy towards anything that appears to need help. For them, it doesn't seem to be a question of deliberately choosing to help. They don't have to make a conscious decision, they just do it. Not all the time and not all of them, but very often.

In the park next to the pond, there is a large, open-air gazebo constructed 100 years ago of heavy tree limbs. Every summer, someone new—a homeless person—lives beneath its large, round Victorian-style roof. Sometimes, it is two or more people. During the very hot summer of 1988, eight men and women lived there. They were young and very, very poor. They had a few blankets and a small grill on which they cooked fish which they had caught.

The North Pond next to the gazebo had been suffering from a major problem for several years. Something was wrong with the old original plumbing. At the south end the drains were clogged and the fill pipe across the pond on the north bank was not working either. The water was stagnant and soon became covered with thick green algae. It was a stifling summer in which the Chicago area's avian population was suffering as a result of botulism poisoning. Toxins were forming in many shallow lakes and ponds. Ducks, fish and other water animals were dying as a result of it. The shallow water in Lincoln Park's North Pond had become a perfect breeding ground for this deadly poison.

There is a layman's term for what was killing the ducks. It is known is "limp neck disease." The phrase is used to describe the first noticeable symptom that results from a progressive deterioration of muscle tissue. The duck's neck weakens and its head falls to one side. Soon, its legs are affected and no longer respond. Unable to swim, the enfeebled bird drifts toward shore, where it eventually dies.

However, for the 12 ducks which I treated: seven mallards, one black duck, one wood duck and three white muscovys (former pet ducks), the outcome would be much happier.

Their heads carefully propped up with soft towels, the birds were taken to the veterinarian where all treatment was prescribed under his supervision. It is illegal to do it any other way.

With illnesses that have no direct cure, it is hoped that by offering nutritionally-sound supportive care that the body's own immune system will function more effectively. At the clinic I learned how to flush the toxins out of their systems by using tube-fed liquid nutrients. This was accompanied by a course of antibiotics felt necessary to treat possible secondary infections that may have occurred during their weakened condition.

Each morning I would get up and go over to the pond looking for sick ducks that drifted to shore during the night. These I would gather up and take home with me. If I found dead ones, I removed their bodies from the water and buried them to somewhat lessen the chance of infecting the remaining healthy ducks still living in the pond. It was puzzling as to why they stayed in such a hostile place. During my walks I found about 40 that had succumbed to the disease.

One morning, the group of homeless people living in the gazebo watched me retrieve a sick female wood duck from the sour-smelling water near where they were fishing. I quickly explained what had happened to the duck and how toxic the pond water had become and I warned them they could become sick, very sick, if they ate those fish. They didn't know what to say but acted as though they would rather be sick than go hungry.

These people were, however, taken by my efforts on behalf of the ducks and one day one of them, a young man, surprised me by suddenly wading out into the pond to retrieve a duck too weak to swim. I felt strongly that it was too dangerous to enter the polluted water and warned them not to go in to get them; but some did it anyway. Their enthusiasm grew with each successful rescue. Once or twice, they had a sick duck in a cardboard box waiting for me when I arrived in the morning. Always, they would address me as "The Duck Man."

Soon I had the equivalent of an intensive care unit in my home. Before going to work each morning, I tube-fed the ducks and administered their prescribed medications. The latter were given by injection into their breast muscles, a procedure that involved placing the duck on its back between my legs, separating the very dense layers of feathers and finding a thick fleshy spot to insert the needle.

It is often assumed that there is no cure for botulism poisoning. It is, however, a fact that with supportive care, afflicted creatures can be saved. In this case, all 12, every one of the ducks rescued, regained their health and were released back to the wild.

Some were so weak that I had to place soft towels on each side of their heads and under their beaks to keep them in an upright position ensuring that they could breathe properly. The weakest ones received a lesser amount of tube-administered food until they showed signs of being strong enough to keep it down. A real concern when tube-feeding is that if the bird is very weak it may aspirate its food and choke to death. So, for a really sick patient, I've found that, at first, less food is better.

I had many chores to do. Feeding and medicating 12 increasingly-active ducks took a special kind of determination. Who got what first and did I get to everybody were problems I had to resolve before I could even consider having my morning coffee. Most days I had things under control and everyone got what they needed. Sometimes, with only minutes left for me to get to work, one or more of them would get loose as I closed their cage doors after feeding. A chaotic chase would follow until they were recaptured.

The culprits were usually mallards. They responded very well to treatment and quickly regained their strength. The ducks would escape by squeezing past my hands, then flapping wildly around the room, swooping under a table or dresser, charging about on the floor as determined to evade me as I was to catch them. I would find myself running around the room sometimes with a hand net trying to nab the ducks so I could get to work. If my neighbors saw or heard me frantically chasing waterfowl around my house, they were kind enough never to mention it.

Such friskiness, however, was a welcome sign of recovery. There were several others. The first of these was that the bird began to eat on its own and process the food pellets I provided without having to go back to tube-feeding. The next thing I watched for was a natural defensiveness and aggressiveness. As the ducks improved, they would hiss and growl whenever I approached. After that failed to intimidate me, the duck would snap at me and sometimes bite with its toothless, but strong, beak. The many bruises on my arms were testimonials to the improving state of health of my guests. Finally, the last stage of recovery involved a duck's eagerness to groom and preen its feathers. This last procedure was the most important of all. Preening involves transferring a waterproofing oil from a gland on its back to each feather. Without having done this a released duck would simply soak up water like a sponge, and thus weighted down, would drown.

A man I knew, an artist who was a nationally-known wood carver, built an aviary for waterfowl study to aid his carving. After learning what he could from the ducks, he determined to release them back into the wild. These birds, during confinement, were fed and sheltered and, thus provided for, did not behave as they would in a natural setting. They neglected to waterproof their feathers. When released into a nearby lake, 19 of them became so water-logged that they very quickly began to sink lower and lower into the water. Not comprehending the danger, they continued to paddle away from shore following an instinct to seek safety out in the middle of the lake. Efforts to coax them back failed and one by one, they drowned. It was a tough lesson to learn for a man who meant them no harm but tried to give them their freedom.

As my duck patients recovered I would carry them, sometimes two at a time, in a cardboard box to the waterfowl area in the Lincoln Park Zoo to be released. The pond there was not full of algae and also had an aerating spray that helped keep the water oxygenated. Sneaking them into the zoo and then releasing them without attracting attention wasn't easy. But that's what I did after each bird had regained its health. Some ducks upon being released would immediately rocket up into the sky and fly at great speed until they were out of sight. Most, however, would hop out of the box and walk slowly to the water, pausing every few steps to turn and look at me as though it were making sure that I wasn't going to chase it with a net and return it to its kennel carrier.

Most of the ducks immediately reverted to their wild ways as they recovered, but one was different. She was a gentle black duck, a species about the size of a mallard, but whose feathers were dark all over. During her stay with me this bird showed no fear nor was she defensively aggressive. She always remained calm and appeared to be trusting of whatever I did, sitting completely still on my lap even when I injected her twice a day with antibiotics.

This black duck, which I treated for three weeks, didn't seem to want to go. At the zoo's waterfowl pond there is a small 2-foot-high stone fence that runs around part of it. I placed her on it, stepped back and waited. Unlike the others, she just stood there quietly watching me. I felt sad, but reached forward and pushed her off the fence toward the water where the other ducks were swimming. She flapped her wings as she touched the ground and then slowly walked to the water's edge. One foot at a time she entered her own world again. I watched while she began to preen herself and exchange sounds with the other ducks. That's when I usually turn away and leave.

How anyone can shoot such a gentle creature and call it fun is beyond comprehension.

Each duck cost me about three weeks of my time and $100 to restore to health and return it back to the wild. If some day I knock on the Pearly Gates and hear a loud quack, I'll know I did the right thing.

NORTH POND DISASTER

In that year of the big die-off, the North Pond had shrunk and become stagnant, a very large puddle of pollution.

That tragic, hot summer of 1988 in the North Pond helped teach me something about how to get results. The pond, many generations ago, had acquired a man-made drain at the south end and a functional fillpipe at the north end. Together, they were meant to serve the purpose of keeping the pond water flowing and fresh.

The epidemic of botulism poisoning taught a lesson that it was not enough to save individual ducks, but that somehow, something had to be done about the frightening situation threatening their lives. In that summer of the big die-off, the North Pond had shrunk and become stagnant, a large puddle of pollution. The drain was not working nor was the rusty fillpipe at the other end.

The fouled pond water became covered with algae, which prevented sunlight from reaching water plants growing beneath the surface. Without sunlight, they could no longer replenish the water with oxygen through the process of photosynthesis. The temperatures that summer didn't help, either. They hovered consistently between 90 and 100 degrees. The pond needed a flow of water from the fillpipe to the drain and it could have used an

aerator to spray water into the air and oxygenate it. It had neither flow nor aeration.

The results were disastrous. Fish were dying in the pond at an alarming rate. Their dead bodies littered the shore by the hundreds. The ducks were also succumbing. I had taken more than 40 dead ones from the pond. My efforts to save a dozen did not get to the root of the problem. They cured some of the ducks, but not the pond. The pollution with its heavy algae was robbing the fish and the birds of any real food source. Flies would land on the carcasses of the dead ducks. There, they laid eggs. These would grow into maggots and float on the surface of the water. And some breeds of ducks, by instinct, eat whatever is on the surface of the water. They ate the maggots, got sick and often died.

Early on, I found a songbird, a blackbird, next to the shore of the pond. It was paralyzed from the neck down. I took it to a veterinarian different from my usual one. She euthanized it. Today I know a lot more about saving birds and would have made a more heroic effort to keep the blackbird alive.

Despite all this, people continued to fish in the pond. Many of these did not speak English. I tried as hard as I could to explain to them how sick they could get if they ate the fish. One kid listened to me and said that he would take the fish home and ask his mother. I was heartsick over such inadequate answers.

Something had to be done and quickly. Time was certainly working against the pond and its life. The dying was going on every day. The city and the park district had to be made to act to save the pond, the lives of its wildlife and possibly even of the humans who used it.

Looking in the phone book under Chicago Park District and the city health department, I called everyone whom I could think of from the custodians and secretaries on up. Whoever answered the phone heard about the North Pond and its pollution from me.

Nor was I reluctant to create a little hysteria. Next to the fillpipe at the north end of the park was a well-used playground for kids. There was also a small, outdoor cafe. Both resulted in children playing along the banks of the pond. With each phone call, I planted the seeds of how serious the problems caused by the pond and its water were.

I focused not only on the problem of birds and fish dying, but also on the threat of botulism to the children playing there. I did not usually get much of a response at first from the person I had contacted. If you can show that the problem affects all of us, and that we are all in the same boat, something can get done, I learned.

I made many, many calls, picking out numbers in the phone book and asking people for the names and numbers of other individuals. By the time I got to the head people, their secretaries and assistants had already called them and notified them about my complaint.

How quickly word reached through the tiers of government! They told me it appeared to be a serious health issue and I agreed.

The next morning the city had a vehicle out at the North Pond and had men taking water samples with buckets. When asked what was going on , they remained very secretive and would say nothing.

The day afterwards, the fun began. I had not been hopeful that the city or park district would respond.

That day, when I arrived at North Pond, I found it completely cordoned off. It was being patrolled by a group of men to keep people away. They identified themselves as off-duty firemen.

As I walked toward the water carrying a small kennel carrier, a pick-up truck came roaring up with its revolving yellow lights on. The driver started to tell me I couldn't go near the pond, but I interrupted him to say that I was going there to rescue sick wildlife, to take them to a veterinarian. The man looked surprised and pleased. He thanked me before driving off. All that week they patrolled and kept the pond off limits while deciding what to do. They also put up a lot of "No fishing" signs.

One day a large construction-type digging machine arrived and the workers began digging up the clogged sewer at the south end. They cleared out the debris and repaired the valve and then filled in where they had been working. Next, they went to the north end and repaired the large flowpipe there, giving it a new valve, allowing fresh water to once again enter the dead pond.

Oh yes, there was a bonus. In the middle of the pond, and for the first time since it was created in 1884, they installed an aerator, a skinny pipe with a special shower-head-like nozzle on the end. The device stands vertically about six feet out of the water.and produces a rain-like spray in all directions for a distance of 15 feet. It operates 365 days a year, refreshing the water in the pond by mixing it with oxygen as it passes through the air. The falling droplets create a slight turbulence around the pipe that helps keep the area ice-free. Moving water doesn't freeze easily. It is a favorite spot for ducks who remain for the winter.

RELEASE

With the cage wire gone, and nothing to restrict its view and movement, a released bird can find the sudden sight of sunshine, wind and swaying trees as traumatic as sudden confinement.

The struggle for daily survival gives life a force found in no other aspect of being. The how-to of this effort needs to be learned and practiced from birth. If once given up, it must be relearned.

Birds, in a cage for a month or more during healing, get used to being cared for, having nutritious food in abundance and being provided warm shelter. They grow accustomed to captivity just as we do, when we accept a job and a paycheck.

Those which survive best in recuperative captivity are the ones that had already learned to adapt to an urban environment that offers the meagerest of food and shelter. For a recovering creature, a catered life in a cage allows time to clean and restore feathers that have acquired a thin, tacky coat of pollution and dirt and to strengthen bones by eating food supplemented with calcium and vitamins. With fresh water provided, a bird can bathe itself daily and, after a few weeks, lose that street look.

A bird in the city can face many food and pollution dangers, some of them as individually disastrous as the Valdez spill was for the wildlife off Alaska.

One example I encountered early one spring

involved a pigeon that lived under the Grand Avenue underpass at Michigan Avenue above where large trucks often parked and left their motors running. The hapless bird had probably been born in the girders above the trucks and it wasn't long before all of its feathers were covered with an accumulation of oily black residue from the diesel motors' exhaust. Oily feathers cannot insulate against the cold. Without dry, clean feathers, a bird will become sick and die.

This poor creature, when I came upon it, was near her end as she sat there on the sidewalk freezing and unable even to move. I picked her up and put her in the inside pocket of my parka, a place that some who know me call "the pigeon pocket."

I took the bird home, where I put her in a clean cage, and gave her nourishing food and vitamins. First, I tube-fed her with special liquid nutrients. This, for me, is standard procedure. Within a week of her starting to take nourishment, the bird was alert and strong. Within a month, she had preened her feathers and washed herself in a small basin of water I had provided and had emerged with pure white feathers. After a couple of more weeks of rest and relaxation, she was released one morning in Lincoln Park.

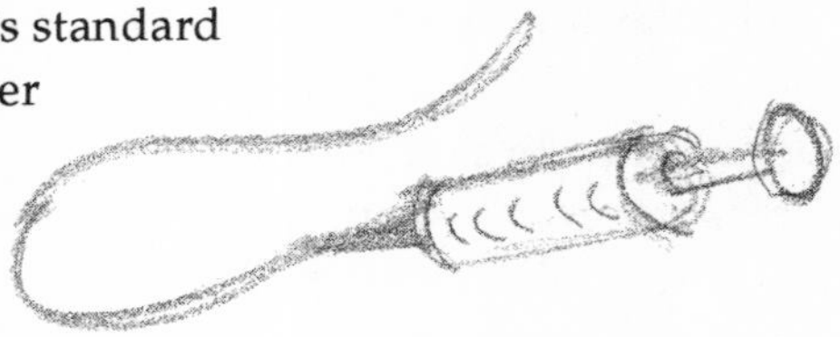

Joyfully, she flew in great circles, far from the diesel smoke that had nearly ended her life.

Nature has no programs of relief or rescue for those unable to care for themselves. There is a moment at which death begins and if relief can be offered even a second or two before that process catches hold, a life can be saved. Collapsed birds very near to death seemingly come back to life and stand up after swallowing just a few drops of honey and water. It was enough nourishment to keep them from the brink. A single day's worth of food or warmth can keep a sick creature alive long enough to restore its health.

RELEASE

During the healing process, birds timidly accept the daily intrusion of human hands changing water dishes and dumping scoops of fresh seed and vitamins. As strength is recovered they begin to defend their small territory with a display of bravado by puffing up and strutting and generally sounding tough whenever their benefactor comes near. Generosity is rewarded with a slap of the wing and a shriek or growl.

While healing is a slow process, the day eventually arrives when the patient is found well enough to be released.

When I feel that a bird is ready to survive on its own, I carry it to the park in a small, unobtrusive cardboard box and search for a quiet area with trees that have craggy branches suitable for roosting. From inside the box, loud sounds of scratching and fluttering indicate an eagerness to be free. Having others of its own kind nearby will make the transition much easier to once again becoming a member of a wild flock . Although I want them out of a cage and back with their friends in the park, it is with feelings of both sadness and cautious optimism that I place the box on the ground, open the the lid and step back. Have I done everything that I could? Are the wings strong enough to allow flight? Did I miss anything? At that dramatic moment, the bird's survival is dependent upon my judgment alone.

After being jostled in a dark container during the walk to the park the bird finds the abrupt exposure to bright daylight momentarily confusing. With the cage wire gone, and nothing to restrict its view and movement, a released bird can find the sudden sight of sunshine, wind and swaying trees as traumatic as sudden confinement.

Most birds, when faced with freedom, act surprised, even shocked. They stretch their necks and nervously give wide-eyed looks in all directions. Not all immediately fly off. Those that don't I have to pick up and hold at arm's length in order to encourage them to fly. And when they do, it often isn't very far. Some flap their wings a few times and land in the grass a few feet away where they cautiously take a few steps forward, always looking back at me. Others, not knowing quite what to do, fly to the closest tree and cling to the bark on the side of the trunk. A few, after being given a gentle toss, surprisingly have flown back to cling to my chest or shoulder! Those that seem truly terrified and refuse all efforts to be released I'll take back home and try to release again a week or so later. Eventually, they decide that being free is not such a bad idea and fly to a nearby tree where they grab onto the closest branch and awkwardly hang on. They may climb to higher limbs and spend fifteen to twenty minutes surveying the surrounding park before disappearing to another tree. Sometimes, after releasing a bird, I'll return a few hours later and find my former house guest still timidly sitting on a branch looking down on birds and squirrels that are busily foraging in the grass nearby. Still well-fed, they seem to puzzle at their fellow park creatures pecking and testing every pebble and seed to see if it is edible. Nothing is passed up. In a wilderness world there is no time to waste, each moment is occupied with food gathering, social interaction and tense watchfulness. Each time I let one go, I wonder how long it will take to regain that sense of urgency.

Dozing Park Workers

The dangerous ice was out toward
he middle of the pond and appeared a kind of dark
blue and looked mushy beneath the surface. Guess
where the bulldozer was heading!

The freezing winters in Chicago are not dreaded by everyone. As soon as the North Pond develops a layer of ice, skaters are seen gliding about its surface both singly and in small groups. Some enjoy playing an unstructured game of ice hockey late into the night beneath the cold light of the lamp posts near the site of the old artesian well. In spite of a loudspeaker's 24-hour repetitive recorded warning of the dangers of thin ice, many young skaters take their chances even when the pond is not entirely frozen over.

To test the ice to make sure that it is thick enough to support the weight of skaters and to make the surface smooth, city workers each winter drive a vehicle with a snow plow attached to the front onto the ice and clear an area about 100 feet in diameter.

One year, a small bulldozer was used.

It was a clear, cold and sunny morning. I was in the park walking my dogs when I noticed a bulldozer chugging out onto the ice from the far north bank. Near the shore where the ice appeared thickest the machine lowered its plow blade and began to sweep the snow aside, leaving the ice smooth.

It was on this very spot some 45 years earlier that my grandfather pointed out to me what he felt was the difference

between thick ice and the type that was too thin to support our weight. He said that the ice that was thickest developed an opaque white color while the dangerous ice was out toward the middle of the pond and appeared a kind of dark blue and looked mushy beneath the surface.

Guess where the bulldozer was heading!

With a sudden loud, cracking crunch, the ice gave way beneath the right rear tread of the vehicle. The driver, no doubt his worst fears realized, jumped to safety just as the back end touched the pond bottom about five feet below the surface. The steel snow plow on the front dug into the ice and kept the bulldozer from submerging completely.

"Ice dangerous! Keep off!" blared the loudspeaker.

The workers ran to their truck on shore and hurried away to get help. They probably thought that if they got the machine out quickly, no one would find out about their blunder.

Meanwhile, a few skaters showed up to try the ice. Astonishingly, some of them skated out to look at the machine, now positioned like a sinking ship. It did not seem to occur to them that the ice might be dangerous for them also.

Thinking it a "good photo opportunity," I returned home and called the photo assignment desk at the newspaper where I worked. The assignment editor agreed and during the few minutes that the park workers were away, a photographer arrived, took a roll of pictures and headed back downtown to develop them. Just a minute or so later, several Chicago Park District trucks appeared and workers quickly began unrolling a long heavy chain. They had no way of knowing that a picture of the sunken dozer was about to be displayed on the front page of the metro section of one of the largest newspapers in the country. What no reader would see, however, was the damage to the park that their attempts to retrieve the machine would cause.

Dozing Park Workers

One of my very favorite trees—one of the park's oldest—was located on the bank of the pond near where the workers' trucks happened to be parked. To the many thousands of visitors who, over the years, entered the park from Roslyn Place, this was no ordinary tree, but rather one of great character whose trunk was nearly three feet in diameter. It appeared almost to be bowing in a gesture of greeting with its twisted old limbs reaching out over the water. Drooping branches adorned with dancing leaves offered shelter to the many water creatures that stopped, for a while at least, to rest and feed at the pond. Many birds and squirrels had made their homes in this tree during its 70 years of existence.

With no other solutions apparent, the workers decided to attach a chain to the bulldozer and then to a winch that was anchored by another chain to the base of the old tree. The motor on the winch was turned on and the chain tightened. The bulldozer did not budge. For a few seconds nothing seemed to happen, then suddenly, the old tree began to fall. It crashed into the water, rolled slightly, then, with its roots torn from the sandy earth, lay still. After surviving nearly three-quarters of a century, this weathered old tree was done in by a chain attached to a hunk of metal in the middle of the pond.

That winter day I watched with anger the tree being ripped from the ground. Not wishing to look anymore, I left the sad scene and went home.

Neighbors later told me how the workers eventually pulled their machine out of the ice then rolled it up on shore and onto a truck.

The next day, the chainsaws committed another insult by cutting up the tree and carting it away. Even though it had fallen, it would have continued as a shelter for the waterfowl, birds and squirrels, and provided a feeling of wildness to the pond.

Where once there was a lush green presence at the Roslyn Place entrance to the park, to me there now exists a gaping hole.

Birds in the Park

Steadying the bird's head with the fingers of my left hand, I grasped the steel shaft of the dart with my right thumb and forefinger.

Many birds in or near the park are accidentally or even intentionally put in danger by virtue of their close living arrangements with humans. In some cases, you can find blowgun darts and lead pellets still in the injured or dead birds.

One afternoon as I walked up my front stairs, a robin flew to the railing on the porch and perched. I noticed a drop of blood slowly forming at the tip of its beak possibly as a result of flying into a window, or worse, a bullet wound in the body. This was a serious injury. The loss of a single drop of blood is life-threatening to a creature weighing only a few ounces. As I moved up the stairs, I decided that a quick grab would be the best method of capture. Leaning toward the robin, I thrust my hand forward as fast as I could. Just missing, I watched as the robin fly off the porch and disappear. I tried to see where it went but could not. You can't help them all, but when it is that close it is sadly frustrating.

It had all started years ago with the bird I named "Patton" and then the ducks.

People should not mistake me and my efforts for those of a wildlife rehabilitator. I'm not one of those. I take wounded animals and birds that I find in the park to an experienced veterinarian and, when they are recovered, release them.

BIRDS IN THE PARK

My four dogs and I go for long walks in the park every day. It has been on these walks that I have had many occasions to find injured birds.

One of these, a pigeon which I found several years ago, was wounded with a three-inch steel wire blowgun dart. It protruded from the bird's skull just above the eye. I heard about its plight from several people who had seen the bird flying around with the rest of the pigeons.

Birds are very wary of dogs and fly off when approached, but this one was different. When I first saw it, the bird was standing quietly by itself in the middle of the concrete expanse in front of the gazebo. It showed no sign of fear as I slowly walked closer. Holding my dogs to one side, I reached down, picked it up, then carried it home.

At home, holding the bird firmly to prevent further injury, I placed it on my lap. Except for the disturbing sight of its injury, the pigeon appeared normal. Although the bird seemed stable, I feared that a trip to the veterinarian might aggravate its injury.

Steadying the bird's head with the fingers of my left hand, I grasped the steel shaft of the dart with my right thumb and forefinger. Gently, I tried to twist it loose but it was stuck fast. It had to come out, so I tried again. Finally, it rotated free. Holding my breath, I carefully withdrew the imbedded projectile expecting to witness profuse bleeding. There was none. There was also no impairment of its ability to walk or fly. The dart must have lodged in a non-critical space between the brain and the eye.

After cleaning the wound and applying an antibiotic ointment, I placed the bird in a cage. Two days later, I noticed a slight discoloring behind the affected eye. By the next day, it was apparent that the eye was collapsing. The tip of the dart must have damaged the rear of the eye after all. I wasn't sure that a bird could fly successfully with only partial vision , so I placed it back into its cage.

He may yet get his freedom for I have recently learned from my veterinarian that a bird with only one eye can survive in the wild.

In another case, on a clear winter day, I went to the park to feed a few remaining ducks that had not yet left the nearly frozen pond. Walking near the bank with a bag of bread I saw a group of seagulls standing on the ground about twenty feet away. One had a red-tasseled pellet gun dart sticking in the center of its white-feathered chest. Intending to grab the injured gull, I slowly walked toward the birds, tossing bits of bread. They all rushed toward me, quickly ate the bread then stood there waiting for more. All except for the wounded bird, who stayed back and just stared at me.

After repeated attempts to get him to come closer failed, I decided to go home and come back the next day to try again. The following morning I returned and discovered that the pond had frozen over solid. Both the ducks and the seagulls were gone. Not wanting to waste a bag of bread, I walked over to Diversey Harbor to a spot near the boat ramp where the water was not yet frozen. There, I found the remaining ducks pathetically swimming in a small open patch of water. A few seagulls were sitting on the dock watching as I approached. At the edge of the concrete embankment I began tossing bread. It was eagerly received by both ducks and gulls. Turning to my right, I saw that standing about 10 feet away was the gull with the dart still in its chest. I was surprised and somehow felt that fate had planned it that way and that I was to catch this unfortunate bird after all. Calmly, I moved toward the seagull. It retreated, however, never letting me get closer than five or six feet. His well-founded suspicion was based upon experience.

Humans can be dangerous. Again, I left him there, hoping the the dart would work its way out, and that eventually he would heal. I returned the next day but the sea bird was gone. The harbor-ice had completely frozen over and even the ducks had moved on.

Often I find baby birds, or even newborn squirrels that have fallen or have been knocked out of a nest. I try to put them back when I can. If their nest has been destroyed, I try to put together a new one. Sometimes the mother will return and care for them; sometimes, not. Baby squirrels I take to a licensed wildlife rehabilitator to raise and release.

One thing I am always concerned about is that when park district employees mow the grass young birds may not be able to escape the cutting machines.

A baby starling I found that way had enough feathers to break its fall, but not enough to fly away to avoid the mower. This bird was raised in my house and, after maturing, was ready to return to the wild.

To help reacclimate it to being outdoors before being released, I put the bird in a large cage suspended outside one of the second-story windows overlooking my backyard. Over a period of days it would again become adjusted to the light and dark that day and night bring, as well as to wind and the sounds of the outdoors. After two days, I wired the cage door open so that, when it felt ready, it could leave and, if necessary, return for food and water. The first day after the door was open the bird refused to come out. During the morning of the second day I happened to glance up toward the cage as I swept the porch. The bird, possibly hearing a familiar call from the park, flew straight out of the cage. It circled the house once, then swooped down at me and gave a loud shrill chirp. I'll always believe it was a thank you and farewell. Flying east, the apparently happy starling sped toward the park. The bird was now going to be able to have at least a summer of life in the park. Perhaps the starling would even reunite with other family members in the tree across the street where it was born.

Dr. Sakas

Birds, for the most part, suffer many
of the same health problems that humans do
and, very often, the treatment is the same.

Dr. Pete Sakas is a grand guy, a veterinarian who gives care from the heart. It seems that every time I have an sick or injured creature he is there, often on Sundays.

Except when my dogs or cats get their annual shots and check-ups, nearly every visit is an emergency situation. I have been there as often as three times a week with sick birds or injured animals, usually from the park.

We generally arrive just before 8 a.m., which is the time he opens his clinic, located in the suburb of Niles. When I spot Dr. Sakas coming down the street I am filled with a sense of relief for the injured creature I have with me. He is a genuine lifesaver.

We exchange greetings, but I can see his eyes focus on the kennel carrier at my side. Inside the clinic, technicians are already turning on computers and other technical equipment which he will use to diagnose illnesses. Their computer-driven analysis machines are astounding. With their ability to analyze each aspect and measurement that can be found in an animal's blood or other bodily fluids they can help a veterinarian find a cure and save a life. They can tell, for example, whether a bird or other animal has been invaded by harmful bacteria or whether its

NILES ANIMAL
7278 MILWAUKEE AVE., NILES, ILL. 60648
PHONE: NILES 7-9325
BLOOD COUNT
PCV_____% TOTAL PROTEIN_____ gm
WBC_____ • WHITE BLOOD CELLS PER 800 RBC
_____/mm³ High Dry Field
DIFFERENTIAL CELL COUNT
Heterophils _____ Band Forms _____
Lymphocytes _____ Eosinophils _____
Monocytes _____ Basophils _____
Thrombocytes: _____
RBC regeneration _____
Blood Parasites _____
FECAL ANALYSIS
Parasites: _____
Microscopic _____
Smear
WBC's

kidneys are diseased or not. The information is nearly instant.

This equipment, I have found, is an absolute necessity to get the quick diagnosis and treatment needed to save a bird or other creature that I have brought in.

To me, observing the healing process has been fascinating. Birds, for the most part, suffer many of the same health problems that humans do and, very often, the treatment is the same. All wild birds, to keep warm and survive a Chicago winter, have to develop a thick undercoat of down feathers. If—due to injury or illness—the bird is unable to maintain this insulation, it may then become chilled, catch cold and possibly develop a respiratory infection.

This infection, when properly identified, can then he treated with proportionate doses of the same drugs that are used in human medicating.

Many illnesses are obvious to a veterinarian and can be treated using known remedies; but sometimes, to find the best medication to combat a stubborn strain of bacteria, the veterinarian must take a culture. This is a smear usually obtained from the bird's throat. It is then placed in the center of a sterile glass dish where a variety of antibiotics are positioned around it like the numbers on a clock. After a period of time, the technician checks to see how the bacteria is affected. If the bacteria grow over an antibiotic it means that that specific one will not kill them. If, however, the area around a sample drug remains clear of bacteria, then this would be an effective choice. The purpose is to find the most specific treatment possible. Usually, however, a diagnosis that is made on the basis of the initial analysis is enough to find a cure. Some infectious diseases need a course of medication requiring daily administration for 10 days while others may require as long as 45 days. When the course is followed diligently, the outcome is almost always assured and the result is that the creature lives.

We humans like to think at times that we are almost divine, an entity completely separate from all other creatures on earth. We are not. All animals get sick, suffering from various cancers, the common cold, infections, gangrene and lung difficulties, to name a few ailments we have in common.

One more sickness animals share with us is food poisoning. For birds and other animals in an urban area, the natural food supplies continue to diminish. Instead of what nature can offer, they eat what people leave behind, from garbage to a half-eaten lunch on a park bench.

Food spoils, but the birds are desperate. They have to eat. Rancid or moldy foods can cause digestive upset for birds. You do what you can for them.

Part of my vision for the future is that there will be more veterinarians who will treat not just privately owned pets and animals, but also those living free in nature, in the city's parks.

The modern city offers pollution, hostility and danger, all of which affects wildlife, and yet, it affords places for animals to live and birds to alight. It, help to work through the incompatibility that results from such a contradiction.

Dr. Sakas and a few wildlife rehabilitators are doing that, but we all need to realize that wildlife in North America, in its full variety, could, with the awareness and concern of its human neighbors, win its battle for a place in our cities. And, yet, as natural resources diminish, it is losing that struggle as surely as it is in Africa and along the Amazon.

The Beaver

One day he just wasn't there anymore.
As mysteriously as he appeared,
he seemed to vanish.

Few sightings of wild animals in Lincoln Park have caused as much excitement as when a beaver was spotted swimming across the middle of the North Pond one winter day. Being a creature more associated with wilderness streams and rivers a beaver inside city limits was unusual indeed.

Where had it come from? As if a foreign alien without papers had been caught, park officials sought a legal ruling on the fate of the woodland visitor, who was unaware of the concern. Lincoln Park Zoo officials were consulted. Some felt that moving the beaver would cause more harm than leaving it alone.

Newspapers printed stories of the beaver's unexplained appearance in which a noted naturalist from a city museum suggested that perhaps it had somehow emerged from Lake Michigan an eighth-mile away. If so, how it managed to safely cross eight lanes of traffic on Lake Shore Drive, and then negotiate Cannon Drive, a busy two-way street, wasn't explained. The only other waterway, the Chicago River was three miles away.

Following its instincts to build a dam, it began to gnaw on small trees along the south bank of the pond. A social animal, its confusion must have been great as it, alone, attempted to duplicate a wilderness habitat inside a city park. Most of the saplings were under two-inches in diameter and many were felled. Each day one could see where the beaver had been during the night by the sight of pointed stakes and short segments of chewed branches where trees once stood. After gnawing through the wood it would then sit on the shore and feed on the bark.

Among park regulars who usually walked around the pond each day the usual greetings and comments about the weather were replaced by, "Seen the beaver today?" Each day we looked forward to seeing him re-create his woodland lifestyle.

One morning as I walked along the east bank I spotted him ten feet away chewing on a twig. This was the closest I had been able to get without him retreating into the water. I became so excited that I pointed out the beaver to a young man, who, judging by the three-piece suit he was wearing, was on his way to work. Not breaking his stride, the gentleman glanced down to where the beaver was sitting then back to the newspaper he had been reading without so much as a raised eyebrow.

To those who visited the park regularly and were familiar with most of its wild inhabitants, the presence of a beaver was a profound mystery. It was to some a hint that this isn't just a world made up of concrete structures and speeding mechanical conveyances. Judging by the smiles on the faces of the people who stopped to look at him, this busy new citizen of the park seemed to represent a bit of the nature that had been pushed aside by the development of the city and surrounding area.

After the initial excitement had worn down I began to look ahead to consider what the beaver's future might be. It couldn't live forever in an area as small as the North Pond. There simply weren't enough young trees to sustain it. Catching and relocating it to one of the beaver colonies located in outlying forest preserves was offered as a solution. The problem was that it was extremely wary and would be difficult to catch. Something needed to be done soon, for its food supply was running out. As people met to decide what to do, the beaver made up its own mind.

As the the supply of nourishing saplings that grew along the bank began to diminish, the park celebrity began to show signs of discontent. A neighbor reported that while walking his dog in the field just east of the North Pond he saw the beaver plodding along in the direction of Lake Michigan. The closest area of water was the boat mooring dock located in the southern section of Diversey Harbor. To get even that far the beaver would have to cross Cannon Drive, the first street east of North Pond. Fearing that it would be in danger from the automobile traffic, my neighbor frightened him into turning back toward the pond. But the beaver was not to be delayed on his journey for very long, nor would anyone witness his departure.

One day he just wasn't there anymore. As mysteriously as he appeared, he seemed to vanish. Each day park regulars went to look for him in the pond and surrounding area but the beaver was gone.

Along the bank I found a piece of a branch he had chewed into a point and took it home.

ICY TALE

. . . once in the frigid water, I had about three minutes before my legs would freeze up. . .

Here is an experience with a rather happy ending that occurred one winter at the North Pond. On this day I decided to stroll down a sidewalk that meanders through the park along the southern edge of the pond. I could catch my bus at the next stop south instead of at my usual one across the street from my house. The pond had a hard glaze on its surface. When the temperature falls below freezing, the pond ices over quickly, except for any place where the water is agitated by low hanging tree branches. As long as the water keeps moving, it does not freeze as quickly as other water around it. Such areas remain ice-free unless the temperature continues to drop. All this sets the stage for what happened that morning.

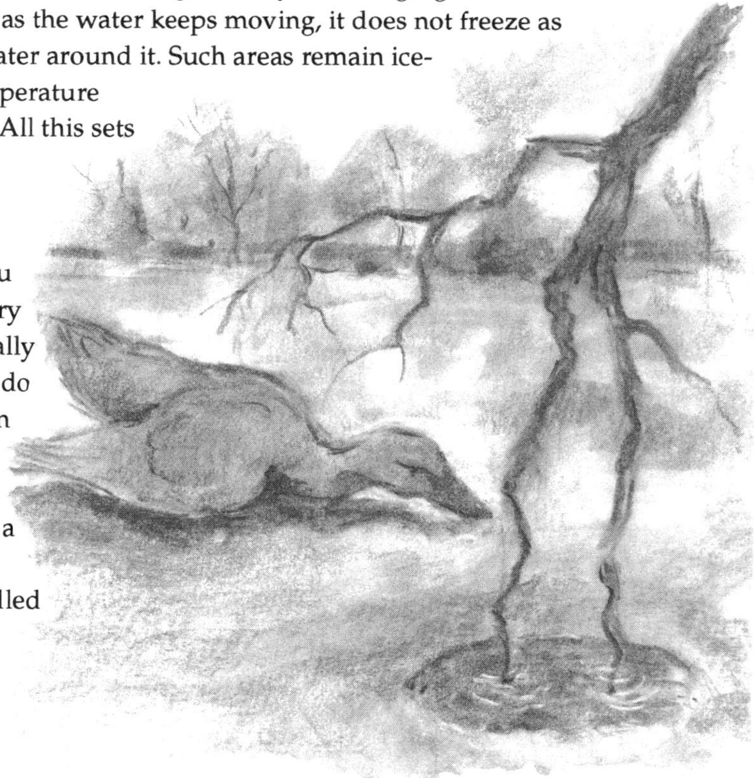

In the wild, you seldom see a solitary bird. They are usually with others. If you do see one alone, it can be an indication of possible injury or illness. That's why a duck, alone out on the pond and huddled beneath a tree branch, caught my eye. It also noticed me and watched with obvious apprehension as I trudged through the snow toward the edge of the frozen pond bank for a closer look. It appeared that during the previous night, the duck, a female mallard, had chosen an unfrozen spot on the pond to sleep. It was a small area located about 15 feet from the bank beneath the hanging branches of a large tree. By the time I happened by and noticed her, she had awakened to find herself in serious trouble.

The bird, her head positioned strangely lower than her body, began to rock strenuously back and forth. I wondered why she didn't fly away as I approached. The mystery was soon solved. The duck was stuck fast to the ice. Apparently sometime during the night, while she slept, the water had frozen and the ice slowly closed in around her. She had awakened to find that her breast feathers were trapping her, having become embedded in the ice. That morning the temperature had remained below freezing and all her efforts could not get her free.

A feeling of panic swept through me as it does when I see a creature in danger. Sympathy and concern for the plight of birds and animals seems a natural part of me, a characteristic I don't fully understand and one occasionally ridiculed by those who are not sensitized to the needs of wildlife in an urban environment. Faced with a life-threatening problem, my mind races through possible solutions as I try to come up with a match of time, apparatus and my ability to effect a rescue. In this case I wanted to rush immediately out across the pond and with a key or some tool chip the ice away from her, but it was too thin to hold my weight. As I continued to watch, she made no progress. I was beginning to be late for work and for a time couldn't decide what to do.

Finally I went back to my house several blocks away to get something with which to chop a path through the ice so that I could wade out to her. It was the only answer I could come up with. In my yard I found a small hand shovel and returned to the pond, where the mallard continued her struggle. Although I was dressed for work, I was determined this bird was not going to be left stranded to freeze. The only thing that stood between her and possible death was a pair of wet socks. There were predators in the park that would be unable to resist making a meal out of a duck unable to protect itself.

I guessed that once in the frigid water, I had about three minutes before my legs would freeze. I had to smash the ice, wade out to the bird, free her and get back to shore in that short of time. The duck watched me very closely as I approached with shovel in hand.

Quickly and with determination, I whacked the ice about two feet out and stepped into the water, shoes and all. The water was so cold it seemed hot as it seeped in where my socks held it in like a sponge. It was most uncomfortable. Reaching farther out I smashed another hole and took another step, lengthening my path to the now nearly-hysterical duck, which undoubtedly thought it was about to be eaten. The duck paused for a few seconds, probably waiting to see if mercifully I would drown, then, seeing that I had not and recoiling from the loud smash of my shovel, she renewed her efforts to escape with even greater vigor than before. As I repeated my blows with the implement and just as I got about halfway to her, she gave a tremendous tug and somehow pulled free, leaving some of her breast feathers stuck in the ice.

Her feet luckily had not frozen to the ice beneath her and consequently when the duck realized that she was indeed free, she began flapping her wings and had no problem getting up into the air. She flew off, probably never so fast in her life, to rejoin her flock on the other side of the pond. I was very happy and I was very cold. My sense of panic for the well being of the duck changed to a profound awareness of what was happening. The water, icy cold, was a foot and a half deep, almost up to my knees. I was also suddenly conscious of the question of how I was going to get back home, change and make it to work on time. Crunching past the people at the bus stop as I walked with my frozen pants, I crossed the street, went home. I changed into warm clothes, and managed to slide into my chair at work as though it were a normal morning.

When—and it isn't often—that I do something like go into an icy pond, people who are close by usually just look away. Seldom has anyone offered to help. The only ones who seem to take any real interest are curious children passing by with their parents and the homeless, who share the park with the displaced wildlife.

PARK MEMORIES

The trees were straighter then: They seemed younger overall and included many more species. Collectively, they constituted a somewhat more supportive woodland home to possum, squirrels, rabbits and pheasants. To me, as a child, the park was a place of adventure and discovery. Most of those trees are gone, replaced by paths and open space.

Several years ago, out in the center of the North Pond, when the Park District decided to dredge it, workers uncovered a well-preserved old-style ornamental cast iron and wood park bench. It was probably used by skaters long ago during one winter and left there as the ice melted. In what has come to seem like the new fashion of doing things, some people brought in a truck and promptly stole this park relic. That bench was one of many testimonials to the age of the park and pond.

I have a picture of my grandfather in the park, sitting on a bicycle. It was taken in the 1890s.

The trees in that picture are more plentiful, and you don't see any litter in the background. The park had the look of a forest, as it does today.

My grandmother told me of those days. She remembered horse-drawn firetrucks racing up and down Clark Street. She also reminisced about how, on warm summer nights, people would bring a blanket into the park, spread it out and sleep on it. There was no fear of crime to keep them from feeling safe and comfortable.

At that time, there was an artesian well at the side of the park, and people came from all over the neighborhood to fill their jugs and buckets with its fresh, clear water. By contrast, the city's main water supply was contaminated by the filth of the stockyards. It would remain so up until the reversal of the Chicago River in the year 1900. My own memories span almost 50 years of the park's history and recall great changes that have taken place in that time. Growing up, I witnessed the encroachment of concrete paths, the removal of the old globe park lights, the thinning of the great, old trees and the development of several grassy areas you can call meadows.

The trees were straighter then: they seemed younger overall and included many more species. Collectively, they constituted a somewhat more supportive woodland home to possum, squirrels, rabbits and pheasants. To me, as a child, the park was a place of adventure and discovery. Almost all of those old trees are gone, replaced by paths, open space, and occasionally, small saplings planted by the city.

I remember when they planted the willows along the shore. They were eight-feet tall and about two-or-three inches in diameter. I have seen them grow into maturity and witnessed many of them die. Some are now a foot thick at the base, and, if they are left alone and remain healthy, may grow as large as the one next to the Conservatory garden, which is five feet in diameter and nearly 50 feet tall. Moving willow branches dipping into the water, help the waterfowl by keep part of the pond unfrozen in the harshest parts of winter.

A magnificent gazebo stood on the ridge on the east side of the pond. It was more than 30 feet across. It was made of logs and overlooked the pond. I remember well the musty smell of the bare wood as you climbed its stairs. It began to decay and park officials must have deemed it cheaper to tear down than to repair. The gazebo was removed in the late 1950s and a large crater was left that is now marked by a loudspeaker attached to the top of a light pole there that in winter blares out the same message 24 hours a day, "Ice dangerous! Keep off!"

FEAST OF THE SEAGULLS

Hundreds of wildly flying seagulls swooped down, sometimes landing on top of each other, vying for chance to tear off a piece of frozen flesh and fly a short distance away to eat it. It was hard to absorb the mad scene before me. What the hell had happened?

With the pond plumbing now operational, all that was needed to keep the water clean and healthy for the fish and migrating birds was to create a flow by turning on the fresh water pipe at the north end and opening the drain at the south end near Fullerton Avenue. After a few days, the valves could be turned off and the pond would have gained a supply of clean water. It was a simple enough idea, but one day during the following winter, someone screwed up.

During a walk around the pond, I noticed that the water level was much lower than usual, although not yet dangerous enough to threaten the pond's creatures' survival. At the south bank a strong current of water flowed into the sewer drain indicating that it had been opened. Because the water fillpipe at the north bank of the pond was left off, the pond was now draining like a ten-acre bathtub. By the time my neighbors and I realized that there was a serious problem, the water was all but gone from the pond.

The following morning as I stepped outside my front door I was startled to hear the shrieks and cries of many seagulls coming from the direction of the park. I hurried across

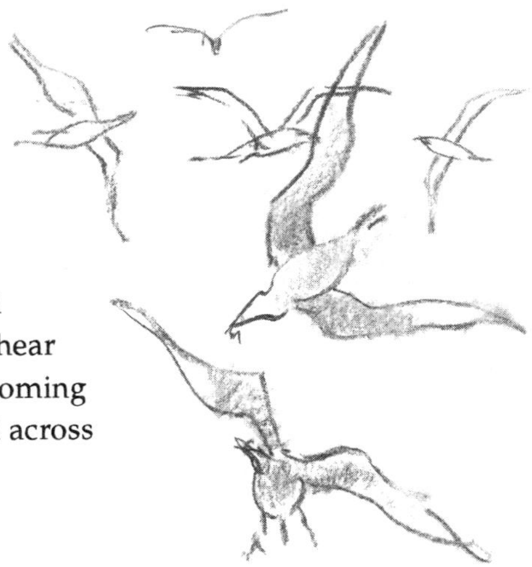

Lake View Avenue and Stockton Drive and stood on the pond's edge. What I saw was incomprehensibly macabre. The entire surface of the now frozen pond was covered with the bloody remains of hundreds of large fish and was alive with frantically feasting seagulls. During the night, the temperature had dropped to below freezing and the water, what was left of it, turned to ice. The level of the pond was by then less than a foot deep and many hundreds of fish froze to death as the ice slowly closed in around them. Almost all were frozen half in and half out of the ice.

Hundreds of wildly flying seagulls swooped down, sometimes landing on top of each other, vying for a chance to tear off a piece of frozen flesh and fly a short distance away to eat it. It was hard to absorb the mad scene before me. What the hell had happened?

Frantic calls to the Chicago Park District brought apologies and an excuse that an inexperienced worker had accidentally left the drain valve open. They promised to rectify the situation as soon as possible.

It was a rare opportunity for the seagulls to get a meal during a time of scarcity in the winter, but for the fish it was a disastrous slaughter. Some of the smaller fish were still alive in still liquid pockets of water near the shore where tree branches agitated the surface preventing its freezing.

The water was eventually restored and the fish replaced, but it illustrated, to me at least, the terrible gap that too often exists between those whose responsibility it is to care for and maintain the park and the real needs of this struggling city wilderness and its creatures.

PARK CONFRONTATION

Spinning around, he began to make pleading, grunting and whining noises as though he were a character in Alfred Hitchcock's movie, "The Birds."

Lincoln Park is a permanent home to many wild birds. On the other hand, excluding a few otherwise homeless individuals, humans in the park are merely visitors. It is a perspective that we easily miss. One young man, however, learned the lesson rather decidedly.

A long, diagonal walkway through a corner of the park connects Lake View Avenue. to a bus stop on Stockton Drive near Fullerton Avenue. The walk is lined sparsely with trees and is illuminated by lamp posts at either end. One of these is situated next to a bench about 30 feet from the Lake View sidewalk.

Late one evening, I was walking my dogs nearby when I witnessed an encounter between the park's wildlife and one of its human visitors.

The lamp posts attract a host

of insects which must number in the many thousands. They hop and fly and buzz around bright lights at the top of the posts. Park birds, night swifts, feed almost exclusively on these bugs.

Like bats, which also munch on insects at night, these birds fly at great speeds through the dark, using long tapered wings as rudders that allow them to twist and turn in flight as they sweep in large circles, keeping their mouths open to catch bugs.

These birds probably kill far more mosquitoes and other insects than all the chemicals ever concocted to do so. It was a while before the experts finally figured out that the chemicals were eliminating not only insects, but also the birds that feed upon them.

These birds—and I've seen one up close at my veterinarian's clinic—have large gaping mouths that they keep open and use to swoop up insects as they fly. They are so unbelievely fast in flight that people are often not certain just what had flown past them.

Enter on this scene a visitor to the park walking down Lake View and heading toward the Arlington bus stop. It was around 11 p.m. and he was carrying two large, heavy-looking suitcases, one in each hand, as though he were preparing to go on a long trip.

He turned down the diagonal sidewalk and was just coming into the light where the swarm of insects were buzzing around the lamppost next to the park bench. These were the intended prey of a single night swift flying in wide, swooping circles as he gobbled them up. This bird darted inadvertently past the head of the visitor, who looked up but was blinded by the light and couldn't tell what had flown past him. The young man, startled, slowed down and proceeded cautiously, twisting his neck to make out the form of whatever it was that now appeared to be stalking him.

Just then, the night swift returned on its arc. The individual, apparently feeling under attack, panicked. He started awkwardly swinging the heavy suitcases around his head like great pendulums. Spinning around, he began to make pleading, grunting and whining noises as though he were a character in Alfred Hitchcock's movie, "The Birds." He ran closer to the light perhaps feeling safer there, but so did the bird, for that's where the food was.

His performance was startling. If there had been loud rock music playing, no one would have thought anything unusual about his crazy dance. But the only music was this poor man's whines and grunts. Passersby on Lake View Avenue stopped to watch. They could not see the bird, which was on the far end of its circle. All they could see was a peculiar young man with two suitcases strangely attacking the air. My dogs sat down and watched, as though not expecting to see this much excitement again for a long time on their nighttime walk.

Again, the night swift returned. This time the young man seemed to be completely convinced that he was being attacked by demons, lost whatever composure he had left, and began running back and forth on rubber legs. Finally, in sheer terror, he spun around and ran back toward the safety of Lake View Avenue, his shrieks disappearing into the night.

The bird flew on unconcerned, as though he had experienced it all before.

TREES

Great Horned Owls and sparrows may sit fluffed up for warmth in the same tree concerned less with each other than surviving through the night.

When I was a small boy, it was not uncommon to see dozens of trees in a two-by-five block area that stood 60 to 70 feet high and that were two to three feet in diameter.

Most of those trees, as they aged, developed wrinkles, crevices and holes deep into their trunks that created perfect homes for a variety of birds and animals from woodpeckers to squirrels and even bats. Some of the best shelters in a tree are created when a branch, torn off during a storm, leaves a slight depression where it has separated from the main trunk. Such a spot, unprotected from the elements by thick bark, soaks up water and begins to decay. Once discovered by birds and squirrels, the pulpy wood in the new hole is hollowed out by pecking and chewing to allow the building of a shelter or nest.

The weather along the lake in Chicago can reach extremes of bitter cold and suffocating humid heat. Winter, sometimes blasting the park with 20 below zero winds shows little concern for the

welfare of creatures trying to find food and shelter. Great Horned Owls and sparrows may sit fluffed up for warmth in the same tree concerned less with each other than surviving through the night. Some robins untypically choose to spend the winter here surviving by seeking warmth in the open areas of the plumbing sheds on the east side of the North Pond. What they eat in winter is anyone's guess. At night, some robins sit near the warm air vents next to Columbus Hospital across the street from the park.

Competition for shelter is fierce. Just as people find that parking spaces become fewer and fewer each year so do the parks creatures, with each fallen tree, have to be lucky to have a place to even temporarily call home. Summer's sometimes 100-plus degree heat can cause those with respiratory ailments much distress and I'm talking about birds and animals as well as humans.

In addition to providing the wildlife in the park with homes and food, the trees also have a special meaning to me as an artist. I look at the shapes and textures of the bark that protects the trees from weather and insects. I notice the way that the branches fan out to form patterns that allows the leaves to catch as much sunlight as possible, for it is the leaves, each a miniature tree in itself, that are the first line of gatherers for the nourishment that allows the tree to live. I especially wonder at the barren weathered tips of the tallest branches. They above all have stories to tell. To me they are the nuggets of the trees. These dried-out and cracked fingers have touched the worst and the best that nature has to offer. They have been witness to generations of creatures over many seasons.

Occasionally, during the fiercest wind and rain storms, the very tops of these branches, swaying in 40 and 50 mile per-hour winds, break off and fall to the ground. I cannot resist picking these pieces up to look at and some I take home.

Twenty years ago, in a three-block area, 29 of these trees were uprooted during a fierce thunderstorm. The next morning I stood among the wreckage humbled by naked roots that stood over ten-feet high. To me it was a tragedy. Then. as now, it was my practice to search through the branches of such fallen trees for birds and animals that might have been injured in the crash. Amazingly, I have yet to find even one. A day or two later, the tree crews dismembered the old giants and trucked away their remains. You can still see, some twenty years later, circular depressions in the earth where some of the largest of the trees once grew.

It was my grandfather, an old Frenchman and oil painter, who introduced me to the beauty of the park's trees. Every day we walked through the park and watched the trees swaying against the sky. We noticed that the trees seldom remained still. It appeared to me that that the action of the wind was as important as water to their growth for it

seemed to allow almost every leaf a chance to get at least some sunlight. We also enjoyed watching clouds, especially if they were small and floating all by themselves.

My grandfather, using brushes that were nearly worn down to the metal bands that held the bristles together, liked to paint pastoral landscapes that sometimes included a tiny solitary figure walking on a distant hill. While I always thought it was me he was putting in his pictures, I now see that it was himself, a rather lonely man, that he was painting on those yellow hills.

After our walks we would find a bench to sit on and he would reach into the breast pocket of his coat and take out a folded piece of white paper that was used to wrap the bread he bought at the corner grocery store. He had a well-used yellow pencil and an old pocket knife with which he taught me how to carve a point without breaking the lead. Looking back and forth from the paper to the trees, he made small drawings, trying one composition, then another. One day, while sketching an old cottonwood, he suddenly put down his pencil, squinted his eyes and said with a tone of amazement, "I never noticed it before but all the tips of the branches reach upward toward the sky. They grow toward the light. I never noticed that before." I was kind of struck that here was something that my grandfather did not know.

These pencil sketches were often preliminary drawings for oil paintings which he did at home on stretched canvases set up on a large wooden easel in a corner of our kitchen.

As a child I always drew and even sculpted animals in clay and wood. When I reached 10 years old, my grandfather began to teach me in earnest his technique of oil painting. There seemed to be an unusual sense of urgency about it. Upon small pieces of shirt cardboard he would sketch trees and meadows and quaint houses set in the woods and I would paint them. He would both criticize and praise my work and most importantly show me how to turn mistakes into positive parts of the painting. He told me about the importance of putting life into my work. My concentration was perfect. I loved doing it. I loved the smell of oil paint and even of the turpentine used to thin the paint.

He died the following year, but those lessons of observation and thought have remained with me for nearly half a century.

Trees

THE FRESH AIR

The group of smokers in the group are all staff members of the hospital. They come out to the park to do the very thing that brought many of the patients into the hospital.

The air in Lincoln Park, long one of its greatest assets, is meant to be clean. Chicago, because of auto emission requirements and having fewer in-city manufacturing plants, is doing a better job than it used to do keeping it that way. Not all those who use the park do so, however. At the north end of the pond, strollers, joggers and park visitors must detour around a huddle of cigarette smokers sitting on the benches there.

These persons seem unaware of the park's long tradition of providing fresh air and of park users' almost magical belief in the benefits of a clean atmosphere that you can actually feel in your lungs. Many people, especially the joggers and bicycle riders, inhale the sweet park air in deep gulps.

Along the lakefront is a structure known in recent

years as the "Theater on the Lake," where patrons in the summer can watch plays and enjoy the lake breeze at the same time. This building was the site 60 years ago of the Chicago Daily News Fresh Air Hospital for the children of Chicago. It was believed then that fresh air could cure just about everything. Youngsters with a variety of serious health problems, especially tuberculosis, were put in this newspaper-sponsored hospital to recuperate. Another children's hospital, LaRabida, still exists along the lake shore in the South Side's Jackson Park.

When I was a child, I went to a day camp adjacent to the large brick and screen structure that now houses the Theater on the Lake. The clean, fresh air off Lake Michigan is obviously considered at a premium for all users of the park, but most especially for the city's children, whether they are visiting the park's playgrounds, swimming on the beach or just romping there with their parents, who hope it can help make kids healthier and stronger.

A saint of the Catholic church, St. Frances Cabrini, had the same idea about the fresh air of Lake Michigan and Lincoln Park. In 1903, she founded Columbus Hospital right next to the park, where it still stands. She herself died there in 1917.

The smokers in the group are all staff members of that hospital. They go over to the park to do the very thing that brought many of their patients into the hospital. You have to wonder how their institution's sainted founder, who was known to have been a pretty feisty nun, would have reacted to their polluting of the park.

ABANDONED CATS

With cat in hand, I rang the doorbell of my dear friend, Mary Jo. I knew it was going to take some convincing and all I asked for was shelter for one night.

My neighborhood next to the park is, to a large extent, one of small apartments with one-year leases. It attracts a lot of single people who, when they move in, often go out first and buy jogging shoes, a walkman and, not infrequently, a kitten.

What motivates people to add a kitten or cat to their household is not always clear. Is it for companionship or because having one is "in"? Either way, many of these residents do not have enough forethought to realize they are making a 15 year commitment to something that is alive, breathing and with emotional needs.

On or about May 1 or October 1—traditional days on which leases end— the area is full of moving vans and rented trucks into which people load that which they consider valuable enough to take to their next apartments. The rest can be seen piled high in neighborhood dumpsters. This includes: tables, chairs, rugs, plants, last year's clothes and, too often, last year's kitten, now a cat.

Like anyone or anything that finds itself homeless and without resources, abandoned cats often go to dumpsters to find food. Instead of sitting in a window absorbing the sun's rays and being quietly passive as in the past, they now find themselves roaming the streets, seldom moving any distance from the area where they were dumped.

There are some people who take these cats in temporarily, but let them roam and the cats are quickly back on the streets again.

The new cat or kitten on the street has no idea where it is or how to survive. It is not unusual that such a lost individual is killed by a car, disease or even by other cats. Some flee the area where they were housed for the park, but are apt to find the competition tougher and survival there even harder.

Either way, they at first try to relate to humans, which is what they were trained to do. The poor little animals can be very pathetic in their abandonment and need.

Late one night, a tiny black cat, perhaps 6 months old, stood in front of a friend's house near the park, crying. Her tail had been broken and was bent at a right angle. She was emaciated and her belly extended. I would later learn that she was carrying five kittens.

The little female was utterly desperate and literally on her last legs. She was following people down the street and whining. A woman walking with a male companion stopped and picked her up. She turned to her date and asked if she could keep it. "Forget it!," he growled. She put the worried little cat back on the sidewalk and moved on.

My neighbor, although he is severely allergic to cats, had taken her in and called me to tell of the cat's plight. I agreed to help and met him half way in the park. Sniffling and tearing, he held the cat out at arm's length as he walked toward me.

With cat in hand, I rang the doorbell of my dear friend, Mary Jo. She greeted me with nervous suspicion. I knew it was going to take some convincing and all I asked for was shelter for one night. We could arrange something else the next day. The cat looked so pathetic, with its bones showing and its extended belly. Mary Jo said, "Yes."

ABANDONED CATS

The next night after work I decided to visit Mary Jo to see if she had reached a decision about keeping the cat. She lives in a small, but cheery apartment. After all, I asked myself, how much trouble could just one little cat be? As I walked toward the park I saw something lying in the middle of the street near the corner. It was a very pretty and very young young calico cat. It was so unused to being outside that it was stretched out in the intersection as though it were on a rug in someone's living room. It was still friendly and cuddly as though it had just recently been abandoned. It was miraculous that it hadn't been hit by a car.

This time, when I knocked on Mary Jo's door, I was absolutely as respectful and as ingratiating as I could be. I had the calico in a kennel carrier behind me on the floor.

Mary Jo, knowing full well how I had wound up with four abandoned dogs myself, had an angry look of "No" all over her face. She agreed, however, to allow it to stay one night. I promised I would come back in the morning and get the cat.

That was eight years ago. The little black cat lost her kittens and we had her spayed. She and the calico have become the best of friends and are still in Mary Jo' s apartment.

Neighborhood cats, if they survive the first few months, try to become hunters and some move to the park, perhaps attracted by the birds and small animals who live there. The ones that have not been spayed, breed. At night, they travel almost the full distance of the pond on the prowl for food, often settling for whatever is dumped in trash barrels. Sometimes park regulars who know the haunts of these misplaced creatures leave pet food for them. Only the toughest become survivors, but whether or not they do, they don't belong in the park.

A LONG LIST OF ABANDONED PETS

They appeared confused and, from their behavior, seemed used to being confined in a small area such as a box or pen.

Incredible though it might be, I have witnessed every kind of pet from a guinea pig to an alligator being dumped in Lincoln Park. The list is long and includes ferrets, ducks, rabbits, ring neck doves and turtles as well as innumerable dogs and cats. Actually, "pet" is the wrong word. It is a misnomer. There is no such thing as a wild creature that is or can be made into a household animal or vice versa.

Following are a few incidents to illustrate.

I was walking my three dogs near the cement casting pier at the south end of the pond. On the east bank, two young men, who had been fishing, were drinking beer and throwing their glass bottles at something in the middle of the water. It turned out to be a small alligator. Apparently someone had determined the animal had become too dangerous to keep and during the night dumped it into the pond.

Obviously, the person had previously thought it O.K. to take in a small alligator and keep it in a bathtub or basement. The individual then decided—also, very irresponsibly—to let it go in the park, believing that the North Pond would make an adequate habitat for the creature.

I walked over to the young men and asked what was going on. They said they had been fishing when they noticed the alligator swimming around. They had tried to catch it with a net and with their hands. The two decided the only way to capture the alligator would be to hit it on the top of its skull with a beer bottle. They could then, they thought, retrieve the stunned animal. Fortunately, they had not be able to hit their target.

At that very moment, two large brown, floppy-eared rabbits loped out of the bushes near the pond and over to where we were standing. They were very tame and obviously also pets that had been abandoned in the park. They were looking for food. The pair was so tame that the two young men kept picking them up and putting them out of their way. I didn't want to leave the rabbits there. They wouldn't have survived more than a day or two at the most. I didn't think that I could carry two rabbits and hold onto my dogs' leashes at the same time so I decided to go home and find a box to carry them in. The two fishermen seemed genuinely bewildered by the situation in which they found themselves with a wary alligator and two confused, but tame rabbits.

At this point a woman came along with a large stroller and two small children. After she said she had cared for pet rabbits before and would take them, one of the young men caught them and gave the rabbits to her. She placed them in a storage part of the stroller and left.

Without a solution to the situation of the alligator in the pond I left and returned home. A day or so later I heard but was unable to confirm that it was eventually captured by others and taken to the Lincoln Park Zoo. Nor could I tell whether it were an alligator or a cayman. A cayman is a small alligator look-alike and is also often thoughtlessly acquired as a pet.

One morning a few days later, a neighbor called and told me that three pet ducks had been dumped along that same part of the bank. I walked over with him and sure enough, we saw a large, white-feathered mother duck with two offspring standing beside her on a short grassy slope near the bank of the pond. They were of a domestic species, the kind some people tend to keep as pets.

They appeared confused and, from their huddling behavior, seemed used to being confined in a small area such as a box or a pen. Traumatically, they had found themselves on the edge of a large pond in a vast city park with unfamiliar sights and sounds and with neither food nor shelter. My first thought was to get them out of the park. Then, I could decide upon a more suitable home.

Trying not to frighten them off, I approached slowly. They were not trusting anyone at the moment. They moved away, crawling under some large bristly bushes near the shore. I retreated a few feet and waited. After I stood there quietly for a few minutes the family of three came out and warily walked toward the water. I hurried forward. As I reached down and scooped up the mother, the youngsters turned and ran back toward the bushes. I quickly handed the duck to a friend standing nearby and dove head-first toward the fleeing fledglings, hoping to grab them before they disappeared beneath the leaves and branches. I landed on my chest and, having stretched my hands as far forward as I could, caught them both just as they merged with the leaves. Fifteen minutes later they were all sitting in my bathtub munching on food pellets and preening their feathers with clean water.

After about a week during which I made sure that they were healthy, I drove them down to the far South Side of Chicago, where I released them in a wooded cemetery which at that time was a safe refuge. On the grounds were a large waterfowl pond complete with feeders and a flowing waterfall. People, grieving the loss of loved ones, found comfort during visits in feeding the ducks and other wildlife. Among the birds that lived there were many species of waterfowl, beautiful peacocks, and large swans. There was even a small herd of deer, some pure white, that roamed the 800-acre cemetery. It truly was, until recently, a Garden of Eden, a park where beautiful living wildlife added a sense of peace to a rather sad place. After new owners "disposed" of all those wonderful creatures, now there is only the call of an occasional bird to remind visitors of what used to live there.

In Lincoln Park, there is no care offered for the domestic and tame animals that people dump. It is not uncommon to see a kennel carrier or a plastic cage, the kind you can buy in a ten cent store, left in the park with the door opened so the pet could "escape."

One day, I saw a white ferret next to a chain link fence near the south end of the pond. Its head was poking out through the wire mesh of the fence and it was looking very bewildered. As I approached, it retreated into the bushes and disappeared. I never saw it again.

Another victim of dumping in the park that I spotted was sitting near the base of a large cottonwood tree next to the water on the west bank. It was a guinea pig, tan and white, a very benign, inoffensive creature. She was standing next to a large plastic bag that was probably used to carry her to the park. I tried to catch her, but she just kept running around the trunk moving faster than I could. An elderly woman sitting nearby must have thought I was crazy as I appeared to run around the tree by myself.

Giving up on catching the guinea pig, I went home and returned with a small hoop net. I propped this at the base of the tree and chased the little creature around and into it. I gently picked her up. She was indeed a female. She also was chortling with incessant tiny squeaks. I went over and explained to the older woman what had what I was doing. She expressed surprise that someone would abandon such a helpless looking creature in the park.

I built the little guinea pig a large habitat in my house. It had hay, vegetables, water, commercial pellets and a wooden A-frame shelter for privacy. She lived five more years and eventually died from cancer.

There is another kind of dumping that goes on. Among certain people in the city, it is a practice to buy ringneck doves—extremely gentle and tame birds—and release them at weddings. These are pet "cage" birds and need to be provided for. They have never been able to survive on their own in the wild.

Their release is supposed to represent love, peace, freedom or whatever. They subsequently are often found starved to death in the park, where they attempt to join flocks of other birds. These doves simply are unaware how to get food or to care for themselves. On a number of occasions, I have found them huddled on the ground, near starvation. After I took them home and cared for them, they flourished.

I have a friend, Erwin Helfer, who is a well-known blues musician and composer in Chicago. Upon returning from a European tour he discovered that one of these doves had fluttered into his backyard. He said it appeared to be too weak or sick to fly and asked if I would come and get it.

The bird looked old, undernourished and in poor health. The more Erwin looked at the dove and talked about its chances for recovery, the more I realized that he was trying to talk himself into giving it a home. A neighbor, knowing Erwin well, had already presented him with a beautiful hand-crafted wooden perch mounted on a floor stand.

Later, I returned with a cage, vitamins and other nutrients for the bird. Erwin, with a comfortable, sunny house, was already explaining the new guest to his two dogs, Corina and Kimbo, both gentle elder citizens who accepted their new companion with an air of disinterested seniority.

He gave the dove the freedom to fly indoors and shopped for a variety of foods with which to feed it. The next time I saw the bird, it was transformed. It was now alert and youthful appearing. Its body was covered with healthy-looking smooth feathers. Feeling at home in its new environment, the dove had begun to sing melodiously. If you called Erwin and got his answering machine, you first heard the bird's voice in the background, then Erwin's message. The dove lets off with a symphony of coos, chortles, notes and melodies that Irwin says rival his favorite composer, Bach.

People tend to fear caring for a creature at first. But, once they do, they can realize what a caring, interesting and rewarding thing it can be. So it was with Erwin and this dove.

Erwin's notation of the melody sung by his
dove on his telephone answering machine

ARRESTS

Knowing that both of them were armed, the policeman, who was in uniform, charged right into the bushes.

Some humans see the park as a place to kill. One summer, I had five individuals arrested in less than a week for shooting at wildlife in the park with pistols and blow guns.

Three of the five arrived together one morning, each armed with a four-foot-long blow gun. These devices consist of a metal tube through which a projectile can be expelled with great force. These were not toys and were capable of shooting three-inch steel darts with enough force to kill just about anything in the park, including people. Blow guns are considered, I am told, gang weapons and possession of one is against the law.

The three young males, about 15 or 16 years old, walked to the water's edge trying to get close shots at the baby ducks. They had been shooting at other birds when they saw that the ducks had moved closer to shore when people started tossing bread into the water. The mothers came first and the little ducklings followed. Furious at what they were

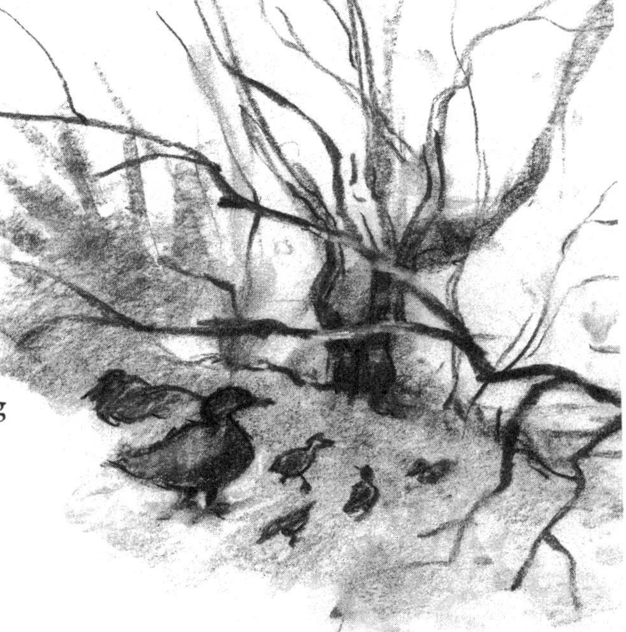

doing, I started looking around for the police. I wanted them arrested. Finding none, I ran home and called 911. When I returned to the pond, the three teenagers were still shooting at the ducks. Anticipating the immediate arrival of the police, I went right up to them and and they stopped. Intimidated, they then walked away toward the kids' playground at the north end of the pond.

By now, 25 minutes had passed and the police hadn't showed up. I went back and called again. I told the dispatcher that these people were shooting at ducks right next to the children's playground.

As soon as I got back to the pond, an unmarked car—part of the tactical unit—arrived with two plainclothes policemen. The police would, I feared, just dismiss the matter and not think the incident was a big deal.

"Maybe if you just take those blowguns away from them," I said.

One of the policemen quickly turned his head toward me, still very upset that no other police had responded for 25 minutes.

"Take 'em away? " he challenged. "They're going to jail!"

His response surprised me.

The offenders were a block away near the playground, looking for something else to shoot.

"C'mon!" said the policeman.

We quickly got into the unmarked car. With me was my German shepherd, Delila. She and I got in the back seat. The police car drove up over the sidewalk and roared across the grass to where the three were standing. The teenagers were caught by surprise. Their blow guns were still in their hands.

After slamming on the brakes, the police flung their doors open and jumped out, radios and nightsticks in hand.

"C'mon, Delila!" I said and we both bounded out a split second behind the two policeman. Even though she was a German shepherd, and may have looked aggressive, she was, in fact, a very sensitive and inoffensive dog. She had a mother's instinct to protect and comfort anyone or anything injured or helpless. For example, I have a photo of an injured duck snuggling up next to her during recuperation. Still, bounding out of that police car, she could have fooled anyone and that day I believe she did.

The teenager closest to me was not a small guy. One of the cops yanked away his blowgun and crunched handcuffs on him.

"Ooo!" the kid exclaimed.

The cop put his face right up next to the teenager's and mockingly said back, "Ooo!"

He then asked, "Whadya think you were doing?"

The teenager answered, "We just wanted to kill the birds."

Delila at first didn't know whether all this was a game or not. Then, I think, she realized it was serious. She just stood there and watched. The kids must have thought she and I were part of the police unit. Also, they were clearly shocked that they had been caught. That was not enough for the policeman. He promptly arrested them. He turned and quietly thanked me and said he would take care of the matter. He put the three teenagers in the back of the police car and drove to the 23rd District lockup.

Later in the week, a similar incident occurred, one that appeared at first was going to be much worse.

It was a sunny morning and I was walking with my dogs along the south end of the pond. Two boys arrived on bikes. One was about 14 years old and quite big. The other was only 8 or 9 and was small.

Each had a pistol tied to the frame of his bike, right under the seat. The guns had been placed in such a way as to hide them. One pistol resembled a .357 Smith & Wesson revolver. The other weapon seemed to be a smaller, a .22 caliber automatic. Actually, both were .22 caliber pellet-shooting replicas. They were gas, or CO^2, operated and had enough force to kill ducks or put out a kid's eye. I had no way of knowing at this time whether or not their guns were real.

The two youngsters rode their bikes into the bushes next to the pond. They put them down and started shooting at the ducks. Each shot made a loud popping noise. I followed them around the pond, at the same time looking for a cop. I wanted them arrested.

There was a police car about a block away. I ran over to it. By now, the boys were back in the bushes, shooting at the ducks.

Knowing that both of them were armed, the policeman, who was in uniform, charged right into the bushes. It was more than a bit tense and that fact was not lost on me. I stayed back.

"Gimme that gun," the cop shouted at the bigger kid.

The two boys sheepishly surrendered their pistols. The older one got mouthy, however, with the cop.

The policemen put them both in his car and drove off. I watched as he stopped about a block away. I thought he was going to give them just a lecture and let them off. I was mad. Instead, the policeman returned, picked up the two kids' bikes and took them with him. By this, I inferred the two were at least going to be taken to the station. I was glad.

For me, it was five arrests for carrying weapons in one week. I couldn't believe it.

A SQUIRREL'S TERROR

One of the most terrifying experiences on earth is knowing that some large creature is about to do everything it can to catch and kill you.

In Lincoln Park, many people seem to enjoy inciting their dogs to attack squirrels. It is a kind of blood lust evidenced in more primitive times as the English "sport" of bull baiting. Then, they would put a bull in an arena and let a pack of especially vicious dogs rip it apart. They did the same thing with chained bears and other large animals, breeding mastiffs especially for this.

The practice of chasing down a fox until it had become too tired to run was little different, because then they would then let the 50 or 60 dogs tear the animal apart.

The remnant of such practices, as barbaric as they may seem, still exists in city parks today. Many think it fun and great sport to unleash their dogs and allow them to chase squirrels. I have found many dead ones in the park, crushed to death with fang marks in their bodies.

I would never permit my dogs to be off their leashes in the park or to chase squirrels or any other animals. These creatures, as it is, have a very difficult time getting enough to eat and are often in a weakened condition. They are dependent on what people give them and on the nuts and seeds from a decreasing number of shelter trees.

One of the most terrifying experiences on earth is knowing that some large creature in about to do everything it can to catch and kill you. Yet many people, when admonished for letting their dogs run a squirrel down and grab it in their jaws, say, "It's O.K. He let go and the squirrel ran up a tree."

What actually has often happened is something similar to what I found on a late evening walk in the park. I saw a squirrel clumsily trying to scoot away. I walked closer. I was shocked to see that it was pulling itself along by its front legs and dragging its hind legs behind it.

Dropping my jacket over the frightened squirrel, I carefully picked it up and carried it home. Over night, I kept him in a kennel carrier with water, food and a folded towel to lie on.

At 6:30 a.m., I got up to take the wounded little animal to the veterinarian before I went to work. He had, however, died during the night. Examining his body, I found two large teeth marks in his abdomen and two in the rear portion of his spine. His back was broken and his pelvis crushed. You could feel the smashed bones in his pelvis move around. They had been fragmented.

This squirrel had died in great pain.

When anything suffers trauma, when there is spinal damage, broken bones, the tearing of muscle or internal injuries, paralysis tends not to be instant unless the spinal cord has been severed. Adrenalin was pumped into the creature's body when it was under attack. This is often enough to give it one last burst of energy and control, to allow it to climb a tree, where the damage takes over. This gives the appearance to the owner of the dog that the squirrel has not been hurt in a major way. However, unable to move, the creature soon falls to the ground where it later dies.

Squirrels in the city face another menace. One morning while walking in the park near a bus stop, I spotted another squirrel dragging itself to a tree and then trying to climb it. She would get up about a foot and then drop back to the ground. Again, using my jacket, I carefully picked her up.

A SQUIRREL'S TERROR

The veterinarian gently took her out of the kennel carrier and placed the squirrel on an examining table. The little creature was very weak and did not resist. Carefully moving his finger tips over her silver-grey fur he finally said, "Here's the problem." With a scalpel he quickly made a small incision in the skin of the abdomen. Squeezing gently, a .22 caliber pellet fell out onto the table. Someone, probably from the safety of their car, called her over, then shot her in the stomach. "Why do people do that?" he said in a pleading tone.

After cleaning the wound and administering various medications to prevent infection, he sensed a good chance for recovery. So much so that I was allowed to take the squirrel home, where I would be able to check on her condition every few hours throughout the night.

At home, I placed her in a small kennel carrier upon several layers of bath towel. Although at first the squirrel seemed to be quietly stable, during the night, she began to cry. The painful wound proved to be more than her small body could handle. Over the next few hours, her cries turned to soft, rhythmic moans. At first light, there were no more sounds.

At 6 a.m., she lay there quietly looking at me with eyes that no longer seemed concerned with worldly things such as fear. Wishing that by just holding her tiny body the death process would somehow be reversed, I gently picked her up and told her how sorry I was. So very sorry. The once wide-eyed little creature whose friendliness and curiosity had gotten her shot, closed her eyes half-way, then died.

I buried the squirrel in the backyard, again apologizing for my species and for whoever made her suffer and killed her for amusement.

Perhaps the assailant was the man my neighbor later saw get out of his car with a .22 caliber rifle and shoot at squirrels and birds in the park near Fullerton Avenue. He got back into his car and started to drive off. My friend wrote down the license number then flagged down a passing police car and reported what he had seen. The police immediately sped after the car and stopped the shooter. With guns drawn, they ordered the man out of his car. After a quick search produced the rifle, they arrested him.

Another time, while walking my dogs, I met a new neighbor, oddly dressed in safari-type clothes, who recently had moved into an expensive penthouse condominium overlooking the park. The way he talked about his dogs, I assumed he was a fellow animal lover. Then, he told me he had spent the morning at a game farm shooting ducks.

Next, he talked about the park's squirrels. He said they were not any different from vermin and had no right to exist because they no longer survived solely on forest food such as nuts and seeds, but instead on what people leave or give to them.

The woodland squirrels, which have been in this area since time immemorial, to him no longer had a right to be here.

Deciding that I knew as much about this individual as I wanted to, I walked away. It was the last time I ever spoke with him.

To me, the park without squirrels would be like the sky without birds. One of the happiest things in the park for me is to watch young squirrels—new to life—romp and frolic around their favorite trees. They are a joy. It is a pleasure to watch them tumble over each other as they run through the grass playing squirrel games that apparently have few rules to get in the way of the fun. I wish everyone felt that way about them, but, unfortunately, some do not.

Just about every squirrel in Lincoln Park has silver-grey fur. However, a few years ago, a squirrel with beautiful black fur was born in the park, possibly the result of a stray gene. It lived near the children's playground and became the mascot of the park. Everyone was thrilled to see it. Everyone fed it.

One afternoon, an elderly woman neighbor phoned to tell me she saw him dead on the sidewalk that runs beneath the Deming

Street bridge just west of the playground. I immediately went over to the park and examined him carefully. He had a bleeding head wound and appeared to have been beaten to death, probably with a stick. He is buried at the base of one of the trees in which he used to frolic.

I never saw a homeless person in the park do that kind of thing. It was someone with a home and with advantages who killed him. Of that, I am certain.

SAWDUST FOR A ROBIN

It is ironic that in 1870 the park commissioners filed a protest with the Board of Public Works against "the inartistic manner" in which the city crews were trimming the trees. And they had only hand-saws!

The park is full of stories, each with its beginning, middle and end. A robin was sitting in a pile of sawdust that had recently been its tree. It had been a beautiful, 80-year-old cottonwood, huge and graceful, one I particularly enjoyed. Did the bird wonder what had happened to the friendly arms that had held its nest?

Although the bark of such a tree may be up to four-inches thick, it affords little protection against the pressing force of 40 and 50 mile-per-hour winds that pound the lakefront during spring and summer storms, some of which are so fierce that rain blows horizontally. The cottonwood had survived nearly a century of battle with Chicago's unpredictable seasons of extreme heat and cold, sunshine and rain. But, finally one night, during a particularly violent thunderstorm, the old tree, its great system of leaves and branches weighed down by thousands of pounds of rainwater, began to lean. Its massive roots, in a tug of war between the earth and the wind, inched their way upward until they could hold no more. With a groan it began to fall, its branches still reaching upward towards the storm that was killing it. This friend to thousands of people in the park and home to countless generations of bird and insect life, crashed to the ground. The 80-year-old giant lay like a wounded animal. Its leaves and outermost branches were the very last part of it to die. In a pure fantasy thought, I wished that, like elephants in the wild, the companion trees nearby, perhaps grown from the same bundle of wind-borne seeds, were able to sense that one of their kind was in great distress and that they could try to push their fallen comrade upright. Once the cottonwood was down on the ground however, there was no hope. The chainsaws and stump-pulverizing machines moved in, operated with a military-like precision and reduce wood and leaves to little more than sawdust. A former life-supporting resource is no more. Less of the air we breath is purified, less wildlife has a chance to nest.

Like all urban park trees, it was not allowed to remain untouched on the ground as it would have been in the woods had it fallen there. Dying in a natural setting, it would have made a contribution for decades as it slowly decayed and fertilized the earth, nurturing future generations of trees and plants. But here, not even a stump was left. It was doubly sad to me for often saplings sprout up as new growth from decaying stumps. There are examples of that in the park where removing the stump had proved too difficult.

However, in this case, the men and machinery were there early the following morning to saw up the logs and chew up the wood.

Sawdust for a Robin

The machine ate the remainder of the stump with its roots and spit out sawdust as though it were a prehistoric predator, its head bobbing up and down as it gnawed the wood. It is ironic that in 1870 the park commissioners filed a protest with the Board of Public Works against "the inartistic manner" in which the city crews were trimming the trees. And they had only hand-saws!

If a tree survives the storm and only a large branch is torn off, it doesn't take the park's birds and squirrels long to vie for the nook created where the limb broke loose. Each wants it as a nest even though it affords only a minimum of protection from the elements. To survive outside, any shelter, no matter how minimal, is not to be passed by.

As the trees in the park have disappeared, birds—including ducks—have had an increasingly difficult time finding places to nest and lay their eggs. One duck in the area would have normally preferred to nest in the hollow of old trees on the bank of North Pond. She gave up the search and chose instead to cross the street and establish herself in a planter in front of a high-rise apartment building. In a few days she returned to the park after being frightened away by gardeners who were cutting the grass and blowing away debris with gasoline-powered tools.

Birds and plants live in harmony in ways we do not always understand. When trees are altered or carted off to serve certain purposes of human urban dwellers, it does not automatically mean the new arrangement serves wildlife.

Somehow, park designers and caretakers just don't understand what the robin does: something is wrong.

THE TURTLE

Whenever he saw an opportunity,
he would vigorously try to escape.
Captivity was not for him.
He wanted out.

It was an escape! With each stretch of his legs he quickly pushed himself forward along the curb. On a residential street two blocks west of the park, a box turtle was heading somewhere fast. Or was it *from* somewhere? Possibly a backyard pet, this five-inch-long hard-shelled creature was in real danger of being run over by mid-day traffic.

As I bent over to pick him up, he withdrew his head and legs then closed the hinged underside completely sealing his body inside a protective natural body armor.

Carrying him in my hand, I walked home and placed the clamped up turtle in a dry ten-gallon glass aquarium until I could decide what to do. I knew absolutely nothing about turtles but reasoned that they must eat vegetable matter. There was a head of lettuce in my refrigerator. I peeled off a few leaves and placed them next to the turtle. He still wouldn't come out of his shell. When I returned a while later, the lettuce was nearly gone, torn off a bit at a time by his sharp, ridge-like upper lip. Turtles have no teeth.

At a bookstore I bought several books on turtles and their food and habitat requirements. Soon I was trying various arrangements to find one that would not only make him happy, but would allow for easy cleaning.

For a while, he lived in a large, clear

plastic clothes storage box with a wire-mesh top. I placed a piece of wood under one side to tilt it, then filled the other end with about an inch of water.

This arrangement worked well, but something other than his immediate needs was bothering me.

Whenever he saw an opportunity, he would vigorously try to escape. Captivity was not for him. He wanted out. It was my feeling that perhaps he could be released in a pond area far out of the city where he might be able to survive on local waterplants and grass.

The North Pond was not a good choice because the turtles there do not fare very well. People fish for them, chase them and find ways to abuse them.

In a northern suburb there is a large man-made botanical garden that has ponds, a large waterfall and vegetation. I drove up there one week-end with my little friend hoping to let him go. But when I discovered that the ponds were concrete-lined and that their smooth, near-vertical sides would not allow the turtle to crawl out onto the bank, I brought him back home.

A few weeks later, I drove out to a rather wild-appearing nature area southwest of the city where I had previously released the mallard duck that I found in Lincoln Park. I was shocked to find the water in the slough to be filled with algae and practically opaque. It didn't seem that either of these areas was a good choice.

So, I'm still looking for an area that will guarantee him adequate food, safety and other turtles. Until then, he will remain with me in a new habitat that I built especially for him. It has a small pool, shelter and lamp for warmth.

I occasionally reflect upon the coincidence that brought me face-to-face with this turtle and wonder what might have happened, had I not happened by.

NIGHT SENTINEL

With a sense of awe and wonder
I stood and watched this traveler, curious
as to where it had been and where
it was going.

After a new-fallen snow the park at night is like a giant sheet of white paper. From the tops of the tallest trees anything that moves across its surface is quickly noticed. Walking with my dogs one winter evening across an open area between Stockton Drive and the North Pond I had the feeling that we were being watched. I turned to make sure my dogs were close by and noticed that Misty, the oldest of the four, was lagging behind the other three walking at my side.

Except for the occasional sound of a passing car or bus, all was still. The sky was a cool bluish-grey and the air was pleasantly cold, but not frigid.

My attention focused upon one particularly tall old-growth tree about thirty feet away whose vertical trunk seemed to be taller than the surrounding trees. My eyes followed the trunk to the top where they stopped cold. Sitting silently, its head tilted downward, a Great Horned Owl, the largest owl in North America with a body length of two-feet, sat watching every move we made. Its head turned to

follow Misty as she meandered nose-down through the snow. She probably looked like any of the other nocturnal creatures that become prey to the owls at night. While she was too big to carry off, I didn't want the owl even to try to attack her. Those talons can do a lot of damage.

This particular owl had been photographed and written about in Chicago newspapers but its whereabouts were not told. It was by chance that I happened by that spot and found where it had claimed its territory. The story said that observers had found the remains of 150 rats at the base of the trees inside the owls territory. Its night vision and muffled flight feathers make it a very effective predator. And the park is a habitat for many creatures, including an abundance of rodents.

With a sense of awe and wonder I stood and watched this traveler, curious as to where it had been and where it was going. Its presence gave one a feeling of being in the north woods. It allowed me a rare moment of fantasy. In another time and place I might be returning from my walk to a cozy cabin and a warm fire instead of the row-house I lived in across the street.

Calling Misty to my side I gave the owl's tree a wide-birth and trudged home feeling special that I had been allowed, for a time at least, to share the night with a traveler from a far-off forest.

Its body motionless, the Great Horned Owl barely turned its head to watch as we slowly walked out of the park leaving him to the privacy of his night-work.

There would undoubtedly be other creatures to walk across the same snow that night but their fate would be far different. They would not marvel and fantasize at the predator above watching and waiting to kill them. More likely they would not be aware that anything was amiss until a moment of stinging pain, then, for a second or two, the terrible realization that their death had come. It happened every night there in the park.

One day the owl moved on, I would like to think, continuing out of the park, out of the city and back to the forests whence it had come.

The Great Horned Owl left me with two thoughts: first, the magnificence of each individual in a species, be it bird or any other animal form; and second, the realization that the area is on a flyway, the wildlife comes and goes.

And so do we.

CHARACTERS

...he tore another page from the book and with continued ceremony, burned it the same way. This continued as long as I watched.Each time he performed the ritual with equal solemnity.

People who use the park have different interpretations of what it's for. I was reminded of that one day as I watched a man perform what I would have to considered to be a very strange ceremony.

He was a young man and he wore glasses. Under his arm were several books and in his hands was a hibachi, a small, portable iron grill. This in itself was not a curious sight, but it was what he planned on cooking that made me stop to watch.

After setting the hibachi down in the center of a large grassy field east of the pond, he carefully arranged the books next to it and stood back up. With a sense of solemn ceremony, he picked up one of the books and opened it. He tore out a single page, carefully put it in the hibachi. His eyes never left the paper. He reached into his pants pocket and took out a match. After lighting it, he reached down and touched it to a corner of the paper. The paper dissolved into an orange flame then turned to ashes.

Bending over, he tore another page from the book and with continued ceremony, burned it the same way. This continued as long as I watched. Each time, he performed the ritual with equal solemnity.

Amazed and somewhat confused, I

watched as every new page was consumed. After a few minute I continued my walk around the pond, then went home.

That night, I had an opportunity to mention what I had witnessed to a neighbor. With a smile on his face he said that he knew this young man. He once saw him doing the same strange thing and asked him about it.

"No, I'm not crazy," the young man told my neighbor. "I just like doing this. I enjoy it."

Then, there are those who see the park as a traditional wilderness, a place in which to hunt their next meal. Their methods, usually more primitive and ridiculous than successful, match neither their determination nor their perseverance. If they did, all the wildlife in the park would have long ago been food on their tables. These are not the wanton individuals who shoot for whatever pleasure or cruelty some feel, but are often people who come from war-torn countries where sympathy for wildlife and even domestic cats and dogs is considered to be a luxury few of them can afford.

For example, one chilly winter morning, as I walked along the sidewalk near the casting pier, I noticed a little old man standing on the bank close to the water's edge. He wore a long muffler, and clutched a long, folded umbrella in his right hand.

He stood perfectly still and stared at something in the water.

Slowly swimming in a small circle of unfrozen water, about ten feet from shore, were two ducks. One had pure white feathers and probably was an abandoned pet , with wings far too small to permit it to fly. The other, which had become its companion, was a wild mallard, that appeared too old to fly south with the flock that had summered at the pond.

An underground discharge pipe from the plumbing shed warmed the pond water where the ducks swam. This small area almost always remained ice-free throughout the winter.

The elderly man stood completely rigid. The ducks were too far for him to reach and he obviously viewed them as something to be eaten and waited for them to come closer so he could hit them with his umbrella. He appeared to be looking for a meal.

The more I watched this odd drama, the more I was convinced that there was no real danger to either duck and I left. Later that chilly afternoon, I returned to the pond. He was still there, standing in the same spot, waiting and watching. So were the ducks.

The next morning I walked over to the pond and found the ducks still swimming back and forth.

On another occasion, I saw a young Asian man crouch down in front of a squirrel in the park. He was moving the fingers of his left hand together as though offering something for the squirrel to eat. Behind his back, he held a rope tied to a brick. The nervous little creature refused to approach this strange acting human. There seemed nothing cruel about this man. He was simply looking for something to eat. Knowing park squirrels, I assured myself that he would not succeed. I was right.

LIFE

When the woman resisted and refused to give up her purse, the man got out of the car and started waving a pistol.

All of life is special. The process of its creation is the most miraculous and pleasurable occurrence on earth, one that has taken all of time to perfect. It is precious. We somehow have to be sensitized to those facts in order to respond to the needs of all creatures including humans. A person, if he or she should break an arm, can, with help, go to a doctor or to a hospital. Other animals, on the other hand, cannot. They either live or they die. To me, life's purpose is connected with not turning away from those with whom we share this existence when they appear to be in genuine peril.

In the park, weakness and peril are not limited to wildlife. The need to help one another is frequent and great.

One extremely windy day in the park, a frail, elderly man was trying to walk along Lake View Avenue on a morning when even young people were having trouble keeping their balance. Unable to stay upright, the older man had to throw himself down on the hood of a parked car to keep from being blown over. People, watching his struggle from their high-rise windows and from the sheltered bus stop, seemed unwilling to do anything but stare. He clutched the car's hood as best he could as the wind continued to blow ferociously.

I hurried over to him and asked where he wanted to go. He told me that he was on his way to visit a friend in nearby Columbus Hospital, a half-block away. I walked him there and returned to the bus stop. A woman now standing at the bus stop, who earlier had been watching him from her second-floor hi-rise window, said to me, "I wondered if anyone were going to help him."

Not everyone sits back and simply wonders whether someone is going to help.

One evening, about 10:30, I was walking with one of my dogs in the park along Lake View Avenue. A man and woman in a banged-up vehicle drove past, but did so very slowly. They were cruising for something other than a parking space, my sixth sense told me. The woman was driving. The car made a slow turn and stopped near the corner where a lady stood with her purse in her hand.

Across the street from where I stood was a large home, where Raul, a young Mexican college student, was watering the lawn.

The woman driver suddenly got out of the vehicle and leaped at the person on the street, grabbing her purse and trying to pull it from her. When the victim resisted and refused to give it up, the man got out of the car and started waving a pistol. Raul, seeing the woman in trouble, instantly started shouting and, despite the gun, ran toward them. Bolted is perhaps a better word. As I ran toward the commotion, a man who turned out to be an off-duty policeman walking in the park also charged across the street toward the scene, as did I.

The thought of personal safety did not seem to be of concern. Faced with this response from all sides, the woman driver got back into the car and immediately sped off, leaving her male companion to flee on foot. He started running west, away from the park, down a side street. The victim, meanwhile, disappeared into the building on the corner.

I ran to my house to call the police. After bounding a half-block I was out of breath and barely able to speak, but reported what I had seen to a patient 911 dispatcher. As I returned to the corner, it seemed that just about every police vehicle in the district was arriving on the scene and from all directions. There is little doubt in my mind that Raul's unselfish reaction saved a woman from injury.

Trailside

Virginia Moe passed away that year after spending a lifetime doing what she wanted most: ministering to birds and animals who were, to her, the only family she ever needed.

What can a person do with an injured bird or animal? The best answer is to call the people at a local veterinary clinic and explain the problem. If it is an emergency, say so and ask if they can help. Some veterinary facilities will take in baby birds and small animals from the wild, but will ask for a donation of a few dollars to help defray the considerable cost of medication and nutrition. However, if the veterinarians there do not have experience with wildlife or simply do not wish to treat or care for your injured creature, ask for a referral to an experienced, federally or state licensed wildlife rehabilitation facility or individual in the area. These are equipped to help medically and, more importantly, have experienced staff who are willing to put in the time and have the patience that these birds and animals require to get well. In the Chicago area there are a number of nature centers and private individuals willing to help care for abandoned nestlings, raccoons or almost anything wild in need. If one veterinarian doesn't know of any, call another and ask.

This is about one such facility, whose small but dedicated staff for many years has taken care of sick and injured wild creatures that were lucky enough to have crossed paths with a caring human being who didn't turn away.

TRAILSIDE

The Trailside Wildlife Museum is located in River Forest, just west of Chicago. Opened in 1932 as a wildlife rehabilitation center by a caring woman by the name of Virginia Moe, it has, in the years since, been the salvation for many thousands of wild birds and animals.

I first heard of Trailside several years ago when I read a newspaper article that said the center had fallen upon hard times as a result of an increasing number of patients but few donations and supplies.

The following Saturday morning, I decided to drive out to River Forest to see for myself.

What I found was a large, well-used and weathered Victorian style home situated on an oversize corner lot with woods and a pond in the back and a long-abandoned garden in front. I parked on the street and walked up to a wooden screen door at the front of the building. I knocked but no one answered so I let myself in. Sitting in the large sunny vestibule was a thin, frail-appearing old woman. She was very quiet and seemed not to notice me as I walked past. Next to her, lying on the floor, was a small comparably old shaggy dog, who moved about whenever she did. Wearing a plain cotton scarf upon her head and an apron around her waist, the elderly woman appeared as though she were about to go to work. Later, I learned that she did just that for more than a half of a century. Now, she left the work of rescue to others and had a permanent residence on the second floor. She had no human family, but I got the feeling that to her it didn't matter as long as she could be near her beloved wildkind.

That weekend there were just two volunteers to handle the endless stream of people stopping by with boxes under their arms asking for help. They brought in everything from a frightened bluejay fledgling that had been bitten by a cat to a green Amazon parrot that the owner no longer wanted. The latter was referred to a veterinary hospital that specialized in pet birds, one that had many clients willing to take the unwanted bird.

Some parents stopped by with their children just to visit, saying that when they themselves were children their parents brought them to Trailside to watch Mrs. Moe mend broken wings and feed orphaned nestlings.

From a room to the left of the main entrance came the sound of loud tapping. Hopping around the corner to greet me was a one-legged duck, her single webbed foot slapping the wooden floor as she made her way into the hall. Stopping a few feet away she wagged her tail feathers and, in a raspy voice, muttered a few scratchy sounds that I chose to interpret as words of welcome. Stepping around her, I wandered through rooms filled with caged squirrels, turtles, songbirds, doves, hawks, seagulls, owls, crows and a fox.

Birds were crowded into long-ago donated cages that were clean, but in need of repair or replacement. There was a hodge-podge of former dime-store parakeet cages and heavy wired cast-offs. They represented what the center, lacking contributions, could afford.

The two volunteers were never without a baby bird or sometimes two in their hands. They had large home-made eyedroppers with which they could feed several hungry babies without having to refill. They worked non-stop all the while I was there, taking in newcomers.

I introduced myself to the volunteers and tried to learn more about the place and what they felt I could contribute. They had to talk to me while placing feeding tubes in the gaping mouths of countless nestlings. A few were gasping for breath and obviously dying. These still gaped for food. The begging look in their eyes was heartbreaking.

I walked from room to room and made mental notes of what

was needed.

The sparrow hawks needed bigger quarters, the crows needed a new cage, the nestlings needed...the robins needed...the doves needed...

A big, strapping man who looked more like an interstate truck driver than a healer of tiny birds and animals introduced himself.

Appalled by the needs of the creatures I saw there, and not wanting to waste his precious time, I told him that, using professional cage building materials, I could build new, easy-to-maintain steel wire cages. I asked him to go around with me and make up a list of what was needed.

When he asked, "How much is it going to cost?"and I answered, "Nothing," he seemed to glow.

He began pointing out excitedly what this creature would need and what that one would want.

I departed with much to do.

On the far South Side of Chicago there is a business that sells professional holding cages and traps, used, unfortunately, in commercial breeding and the capture of birds and animals. It also sells materials with which an individual can make his own. Over the next few weeks I purchased steel cage wire in hundred-foot rolls, latches, assembly tools and sheet steel to make floor pans as well as watering and food dishes.

The woman there who helped me with my orders seemed curious about what I was doing and when I explained that I was donating cages to the Trailside Museum she offered to help. She said that they had an amount of non-saleable, returned and damaged merchandise that maybe we could use. She went up and down the aisles collecting bits and pieces of disassembled heaters, cage material, waterers, parts of feeders and odd hardware. Each time that I came to buy more cage building material she would have a pile of cast-offs waiting by the door. In all, we must have filled my station wagon twice with donated hardware. Just about all of it was put to good use in one way or another.

From these materials I was able to construct 20 sturdy, welded-wire cages. Most of them were two-feet square but some were three feet wide and a couple were five feet high and final assembly had to be done at the center. They were designed with large, latched doors at the bottom that allowed easy removal of the metal floor pans that I made for cleaning. Access doors and latches were installed at a suitable height in the front of the cages for feeding and watering. Round wooden rods were set at various heights for perches and each cage had numerous water and feeding dishes. I was pleased with them, not only for my workmanship but in knowing that it would help make a better recovery environment for the wildlife at the center.

That year, the president of the Cook County Board challenged the center's right to care for animals. Hundreds of the Trailside's supporters, however, rallied to the cause and marched in protest. After a long and emotional court fight with government wildlife agents actually walking out the front door with scores of injured birds and animals that they intended to euthanize, Trailside won and the creatures were ordered returned by a federal judge. After a new board president was elected, Trailside's fortunes changed for the better with money approved for construction of a small addition to the center and additional staff.

Virginia Moe passed away that year having spent a lifetime doing what she wanted most: ministering to birds and animals who, to her, were the only family she ever needed.

BEGGAR

With part of the money he begged, he bought loaves of bread and bird seed for the other, hungrier creatures in the park. Apparently, as he saw it, they were, after all, engaged in the same daily struggle that he was.

Among the most fascinating and surprising animals in any city park are its human visitors, an occasional one of whom takes up residency there.

This man showed up in the spring. He was one of many in a line of people to live in the park. He was big, unkempt, dirty and sometimes, loud. He worked Clark Street, begging for money, and then returning to the park. Otherwise, he mostly kept to himself. I avoided him, not wanting any kind of confrontation. He was, by his very size, intimidating. I never saw him use his strength or try to threaten anyone, but he was big enough to evoke a reaction without even making the effort. And the man was dirty. His face, his hair, his body were dirty all over. He was not someone I wanted to be near and I found ways to stay clear of his path.

He hunkered down on Clark Street and begged. I often saw him do it, but he didn't seem to get many donations. There have been others like him over the years. They sleep on benches or under them. The police protect us from them. They move them along, keep an eye out for any violations. And just when the men, and sometimes the women start to become familiar, well, you don't see them anymore. You see someone else in their places.

Beggar

This man—a dirty, young, homeless beggar, a big, and by that fact, dangerous-seeming human being, did something I didn't notice at first. With part of the money he begged, he bought loaves of bread and bird seed for the other, hungrier creatures in the park. Apparently, as he saw it, they were, after all, engaged in the same daily struggle that he was.Then he, too, disappeared.

Several months later, he came back to the park. I was sitting and sketching when I saw him. He was shaven and appeared clean, with fresh clothes. He walked through the grass toward the pond in a head-down, shoulders back, determined manner, past sunbathers who pretended that he was invisible. He carried two brown paper bags full of large, round loaves of bread

As he neared the bank, a flurry of wings appeared around him. First one blurred by, then two, then fifty or sixty birds swirled around this man as he opened the bags and began breaking and tossing the bread onto the grass and into the water where hungry ducks, experienced in city ways, quickly gathered. With his long hair he looked like a modern day St. Francis.

People walking by came over the the pond to watch the mallards aggressively tear at the soggy bread and to see the pigeons cooing and flapping their wings trying to stay airborne yet be ready to drop to the ground in an instant to grab what they could. Swooping in and out of the growing gathering of families with children were the tiny sparrows, eager and determined, who seemed grateful to be able to pick up and fly off with just a crumb.

The crowd delighted in watching him pass out the bread. They took great pleasure in it, laughing and picking up stray chunks and throwing it to where they felt the birds would be able to get it.

It was amazing. One could not help but reflect how much life and spirit he had given these people.

I decided to photograph him but as I adjusted my camera I looked up again to see him and he was gone. I could not figure how or where he went. He just wasn't there.

The rest of the people were. They were kept busy passing out the bread and enjoying themselves. The man had created something magical in the crowd.